WITHDRAWN
UTSA LIBRARIES

RENEWALS 458-4574

Magneto-Resistive and Spin Valve Heads

FUNDAMENTALS AND APPLICATIONS

Second Edition

Academic Press Series in Electromagnetism

Edited by
ISAAK MAYERGOYZ, UNIVERSITY OF MARYLAND,
COLLEGE PARK, MARYLAND

This volume in the Academic Press Electromagnetism series is the second edition of the book on magneto-resistive heads. This book is written by John C. Mallinson, who is one of the leading experts worldwide in the area of magnetic recording. He is well known and highly regarded in the magnetics community for his many important contributions to the field of magnetic data storage as well, as for his well-written and influential books on magnetic recording.

This second edition of his previous book on magneto-resistive heads reflects the phenomenal and explosive progress in magnetic storage technology where anisotropic magneto-resistive heads were quickly displaced by giant magneto-resistive heads, spin valves. Naturally, this edition contains many newly written chapters related to giant magneto-resistive heads. These chapters cover such new topics as synthetic antiferromagnet and ferrimagnet multilayer technology, magnetization fluctuation noise in small giant magneto-resistive heads, colossal magneto-resistance and spin dependent tunneling phenomena and their relevance to the future development of magnetic recording.

This is a short and concise book that nevertheless covers an extraordinary amount of technical information. The salient features of this book are its clarity, brevity, straightforward and "nonmathematical" manner of presentation with strong emphasis on underlying physics, unique historical perspectives on the field of magnetic recording.

I maintain that this book will be very attractive as a clear and concise exposition of magneto-resistive heads. As such, it will be a valuable reference for beginners and practitioners in the field. Electrical and material engineers, applied physicists, experienced developers of magnetic recording systems and inquiring graduate students will all find this book very informative.

Books Published in the Series
Giorgio Bertotti, *Hysteresis in Magnetism: For Physicists, Material Scientists, and Engineers*
Scipione Bobbio, *Electrodynamics of Materials: Forces, Stresses, and Energies in Solids and Fluids*
Alain Bossavit, *Computational Electromagnetism: Variational Formulations, Complementarity, Edge Elements*
M. V. K. Chari and S. J. Salon, *Numerical Methods in Electromagnetism*
Edward P. Furlani, *Permanent Magnet and Electromechanical Devices: Materials, Analysis, and Applications*
Göran Engdahl, *Handbook of Giant Magnetostrictive Materials*
Vadim Kuperman, *Magnetic Resonance Imaging: Physical Principles and Applications*
Isaak Mayergoyz, *Nonlinear Diffusion of Electromagnetic Fields*
Giovanni Miano and Antonio Maffucci, *Transmission Lines and Lumped Circuits*
Shan X. Wang and Alexander M. Taratorin, *Magnetic Information Storage Technology*

Magneto-Resistive and Spin Valve Heads

FUNDAMENTALS AND APPLICATIONS

Second Edition

JOHN C. MALLINSON
Mallinson Magnetics, Inc.
Belmont, California

ACADEMIC PRESS
A Harcourt Science and Technology Company
San Diego   New York   Boston   London
Sydney   Tokyo   Toronto

This book is printed on acid-free paper. ∞

COPYRIGHT © 2002, 1996 BY ACADEMIC PRESS

All rights reserved.
No part of this publication may be reproduced or transmitted in any form or by any means, electronic or mechanical, including photocopy, recording, or any information storage and retrieval system, without permission in writing from the publisher.

Requests for permission to make copies of any part of the work should be mailed to the following address: Permissions Department, Harcourt, Inc., 6277 Sea Harbor Drive, Orlando, Florida, 32887-6777.

ACADEMIC PRESS
A Harcourt Science and Technology Company
525 B Street, Suite 1900, San Diego, CA 92101-4495, USA
http://www.academicpress.com

ACADEMIC PRESS
Harcourt Place, 32 Jamestown Road, London, NW1 7BY, UK
http://www.academicpress.com

Library of Congress Catalog Card Number: 2001091917
International Standard Book Number: 0-12-466627-2

PRINTED IN THE UNITED STATES OF AMERICA
01 02 03 04 05 06 IP 9 8 7 6 5 4 3 2 1

*This book is dedicated to my wife, Phebe,
who once again inspired me to put pencil to paper
and did all the word processing.*

Contents

Preface to First Edition   xiii

Preface to Second Edition   xvii

CHAPTER 1

B, H, and M Fields   1

    Magnetic Field H and Magnetic Moment μ 1
    Electron Spin   3
    Exchange Coupling   5
    Magneto-Crystalline Anisotropy K 6
    Magnetization M 7
    Magnetic Poles and Demagnetizing Fields   8
    Flux Density B and Flux ϕ 11

CHAPTER 2

The Writing Process   14

    Write-Head Field   14
    Written Magnetization Transition $M(x)$ 17

CHAPTER 3

The Reading Process   21

    Recording Medium Fringing Fields   21
    Read-Head Flux   22
    Output Voltage   23
    Output Spectrum   24
    Digital Output Pulse   24

CHAPTER 4

The Anisotropic Magneto-Resistive Effect   27

    The Basic Effect   27
    Magneto-Resistive Sensors or Elements   28
    Magneto-Resistive Coefficient $\Delta\rho/\rho_0$ 34
    The Unique Properties of Permalloy   35

CHAPTER 5

The Giant Magneto-Resistive Effect   39

    Superlattices   39
    The Physics of the GMR Effect   43
    Granular GMR Materials   47

CHAPTER 6

Vertical Biasing Techniques   50

    Permanent Magnet Bias   50
    Current Bias   51
    Soft Adjacent Layer Bias   53
    Double Magneto-Resistive Element   54
    Self-Bias   55
    Exchange Bias   56
    The Barber-Pole Scheme   58
    The Split-Element Scheme   59
    The Servo-Bias Scheme   60

Contents

CHAPTER 7

Horizontal Biasing Techniques   64
- Conductor Overcoat   65
- Exchange Tabs   66
- Hard Films   67
- Closed Flux   68
- MRH Initialization Procedures   68

CHAPTER 8

Hunt's Unshielded Horizontal and Vertical Anisotropic Magneto-Resistive Heads   72
- The Unshielded Horizontal Head   74
- The Unshielded Vertical Head   76
- Single Magnetization Transition Pulse Shapes   78
- Side-Reading Asymmetry   79

CHAPTER 9

Single-Element Shielded Vertical Magneto-Resistive Heads   81
- The Function of the Shields   81
- The MRE Shield Magnetic Transmission Line   84
- Shielded MRH Output Voltage Spectra   87
- Shielded MRH Designs   91

CHAPTER 10

Simple Spin Valve Giant Magneto-Resistive Heads   94
- Spin Valve GMRHs   94

CHAPTER 11

Enhanced Spin Valves   100
- Top and Bottom Spin Valves   100
- Exchange Coupling Phenomena   101
- Cobalt "Dusting" and CoFe Alloys   105

Spin Valve Biasing Effects   106
Unpinning Phenomena   109

CHAPTER 12

Synthetic Antiferro- and Ferrimagnets   111
In the Pinned Layer   111
In the Free Layer   115

CHAPTER 13

Antiferromagnet Upsets   118

CHAPTER 14

Flux-Guide and Yoke-Type Magneto-Resistive Heads   121
Flux-Guide MRHs   122
Yoke-Type MRHs   123

CHAPTER 15

Double-Element Magneto-Resistive Heads   125
Classification of Double-Element MRHs   126
Advantages of Double-Element MRHs   129
General Comments   130

CHAPTER 16

Comparison of Shielded Magneto-Resistive Head and Inductive Head Outputs   132
Small-Signal Voltage Spectral Ratio   132
Large-Signal Voltage Comparison   134

CHAPTER 17

Simplified Design of a Shielded Anisotropic Magneto-Resistive Head   138
The Written Magnetization Transition   138
Inductive and Shielded MRH Output Pulses   140
The MRE Thickness T 141
The MRE Depth D 142

The Measuring Current I 143
The Isolated Transition Peak Output Voltage δV 144
Optmized MRH versus Inductive Head Voltages 145

CHAPTER 18

Simplified Design of a Shielded Giant Magneto-Resistive Head 147

The Written Magnetization Transition 147
Inductive and Shielded GMRH Output Pulses 149
The MRE Thickness T 150
The MRE Depth D 151
The Measuring Current I 152
The Isolated Transition Peak Output Voltage δV 152
Future Projections 153

CHAPTER 19

Read Amplifiers and Signal-to-Noise Ratios 157

MRH Read Amplifiers 157
Signal-to-MRH Noise Ratios 159
Magnetization Fluctuation Noise 161
Conclusions 165

CHAPTER 20

Colossal Magneto-Resistance and Electron Spin Tunneling Heads 166

Colossal Magneto-Resistance 166
Magnetic Tunneling Junctions 168

CHAPTER 21

Electrostatic Discharge Phenomena 171

Appendix

Cgs-emu and MKS-SI (Rationalized) Units 174

Recommended Bibliography on Magnetic Recording 176

Index 179

Preface to First Edition

In 1968, the Ampex Corporation Research Department hired a recently graduated doctoral student from the Massachusetts Institute of Technology named Robert P. Hunt. His initial assignment was quite simple: "Find something new and useful in magnetic storage technology." Shortly thereafter, Bob Hunt invented the magneto-resistive head (MRH).

I had the privilege of sharing an office with Bob Hunt and not only witnessed his invention but also provided the first analysis of the output voltage spectrum for both horizontal and vertical unshielded magneto-resistive heads. I have followed MR technology closely in the ensuing 27 years and this book is the result.

Of greater importance is the fact that the office immediately adjacent to mine at Ampex Research was occupied by Irving Wolf, the inventor of the "Wolf" permalloy electroplating bath. When asked by Bob Hunt if he could fabricate a magneto-resistive head, Irving Wolf replied, "Certainly, I'll make it out of evaporated permalloy." This was, indeed, a remarkable piece of serendipity. It is quite possible that, had Bob Hunt and Irv Wolf not been in such close proximity, there would be no MRHs even today. Not only did the first MRHs use permalloy as the sensor, but all MRHs manufactured in large quantities to this day also use permalloy. Moreover, it is likely that most of the advanced MRH designs now being proposed will also use permalloy.

The intent of this book is to introduce the reader to the principal developments in MRH technology that have occurred since Bob Hunt's initial work. To make this book self-contained, Chapter 1 contains a review of the basics of magnetic materials and magnetism and Chapters 2 and 3 cover the writing process and the usual inductive reading process in magnetic recording, respectively. The anisotropic magneto-resistive effect and the unique properties of permalloy are discussed in Chapter 4.

To achieve linear operation with low harmonic distortion, the MR sensor must be magnetically biased with a vertical bias field. Additionally, a horizontal bias field is usually applied to keep the magnetic state of the MR sensor stable. The principal techniques for producing these biasing fields are the subjects of Chapters 5 and 6.

In Chapter 7, the output voltage spectra and isolated written transition output pulse shapes of Hunt's horizontal and vertical MRHs are reviewed. A similar analysis of the very commonly used "shielded" MRH follows in Chapter 8. Here, the important fact is demonstrated that the output voltage spectral shape and isolated pulse shape of the shielded MRH and an ordinary inductive read head are almost identical.

Alternative designs for MRHs in which the MR element is incorporated into the structure of thin-film ring heads are considered in Chapter 9. These types of MRHs are often called "yoke-type" or "flux-guide" designs.

Considerable interest exists in the performance characteristics of MRHs which have two MR sensors. Depending on the directions of magnetization, the current, and the external voltage sensing connections to these double-element heads, they are called a variety of rather confusing names such as gradiometer, dual-stripe, and dual-magneto-resistive heads. These differences and their relative performance advantages and disadvantages are the topic of Chapter 10.

In Chapter 11, the output voltages of a single-element shielded MRH and an inductive read head are studied in ways that lead to a particularly simple and direct way of comparing their performance.

An entirely new physical phenomenon called the giant magneto-resistive (GMR) effect was discovered in 1987. In Chapter 12, the basic physics of this phenomenon is outlined and, in Chapter 13, a proposed design for some giant magneto-resistive heads (GMRHs) is discussed. It is expected that GMRHs will produce output voltages greater by perhaps a factor of 5 than those of conventional anisotropic MRHs.

The penultimate chapter contains a step-by-step simplified design sequence for a single-element shielded MRH. This material is included because it demonstrates in a straightforward fashion precisely which physical phenomenon controls each of the principal dimensions of the MR sensor and the MR sensor-shield "half-gaps."

The last chapter is devoted to system considerations such as read amplifier designs and the signal-to-noise ratio of MRHs. Finally, conclusions concerning the significance of MRHs in future high-density digital recorders are offered.

An appendix which contains a listing of the defining equations and a table of conversion factor for cgs–emu and MKS-SI magnetic units is provided. Despite the fact that virtually all physicists and electrical engineers in the world are trained to use MKS-SI, the magnetic storage industries in the United States and Japan continue to use cgs–emu, and this convention is followed in this book. In my opinion, the preference for one convention over another is an unimportant matter of taste.

Most of the material in this book is presented in a nonmathematical manner. Thus there are no mathematical analyses or derivations. On the other hand, the results of such analyses and derivations are used extensively. My belief is that most readers are not interested in detailed mathematical formalism and often find that copious mathematical detail often acts to obscure the intuitive obviousness of the underlying physics. The lack of mathematics should not, however, lead the reader to suppose that this is an elementary exposition. On the contrary, the aim is to lead the reader, in a straightforward fashion, to a high level of scientific understanding of MR materials and heads.

Finally, my hope is that this book will prove enjoyable and fascinating to read. Even after spending the last 35 years in magnetic recording research and theory, I find the seemingly unending inventiveness of my industrial colleagues both amazing and glorious. I hope that this book helps others to share in this rich and fascinating field.

Preface to Second Edition

When I wrote the first edition of this book in 1994, there was no reason to suppose that anisotropic magneto-resistive heads (AMRHs) would become obsolete in the near future. By 1997, however, giant magnetoresistive heads (GMRHs) were beginning to be installed in hard-disk drives and by 1998 essentially all disk drives used GMRHs. The GMRH was invented, in the early 1990s, at IBM Research and the name "spin valve" was soon adopted, presumably in analogy with the British nomenclature for an electronic or vacuum tube, a valve. Spin valves displaced AMRHs because, with little added complexity, they produce output voltages about 4–5 times greater.

The predominance of GMR spin valves has rendered the first edition of the book out-of-date because it contained but two short chapters on the subject. This second edition redresses that deficiency by including seven chapters of new material devoted exclusively to GMR spin valves. Almost all the AMRH material found in the first edition has been retained, but it has been extensively revised and edited in order to make it relevant to the present-day applications of AMRHs to digital tape recording.

The book commences, as before, with a chapter on the fundamentals of magnetism and magnetic materials. This is followed by a review of the basics of the writing process in Chapter 2 and the inductive reading process in Chapter 3. The anisotropic magneto-resistive effect, first discovered in 1857, is introduced in Chapter 4. The giant magneto-resistive effect, first discovered as recently as 1987, is treated in great detail in Chapter 5.

Chapters 6 and 7 cover all the basic methods that can be used to produce the vertical (transverse) and horizontal (hard) magnetic biasing fields, respectively. Bias fields are necessary for the proper operation of both AMRHs and GMRHs. The original unshielded horizontal and vertical AMRHs invented at Ampex in 1969 are analyzed in Chapter 8.

All magneto-resistive heads are operated today between shields of high-permeability material in order to achieve narrower output pulses. The behavior of shielded single magneto-resistive heads is treated in Chapter 9.

Chapter 10 introduces the basic structure of a simple spin valve and analyzes its giant magneto-resistance versus magnetic field characteristics. All new material is presented in Chapter 11, where many techniques used to enhance the performance of spin valves are discussed. The impressive-sounding synthetic antiferromagnet and ferrimagnet multilayer technology is explained in Chapter 12. This is followed, in Chapter 13, by a review of the phenomenon of antiferromagnet upsets and the strategy employed in a disk drive to recover from an upset. Not only is all the material in Chapters 11, 12, and 13 new, but it also was unknown at the time of the first edition.

The material in Chapter 15, flux-guide and yoke-type MR heads, will be principally of interest to practitioners of digital tape recording and is essentially unchanged from the first edition. Similarly, the subject of double-element MRHs, which can be either AMR or GMR, is covered in Chapter 16. The principal attraction of double-element heads is the suppression of the so-called thermal asperities (TAs) that are due to rapid fluctuations in the magneto-resistive sensor film's temperature caused by magnetic recording media proximity effects.

A comparison of both AMR and GMR shielded and inductive reading heads is undertaken in Chapter 16. Both the small signal sensitivities and the peak digital output pulse voltages are treated.

Simplified design exercises for a shielded AMRH and a shielded GMRH are the subject of Chapters 17 and 18, respectively. These design exercises will be found instructive because they demonstrate how the chosen cgs-emu units used throughout the book work, because they show what physics determines the principal dimensions and performance of the AMR and GMR heads, and finally, because they show that in order to attain optimum performance for an MRH it must be designed optimally to match the transitions written in the recording medium. A forecast of GMR spin valve design from today's 10–20 gigabit/in.^{2} heads to those projected to be suitable for 100 gigabit/in.^{2} is included.

Chapter 19 is devoted to a review of the two types of reading amplifiers that are used and to an analysis of the signal-to-noise ratio of MRHs. In the increasing small GMRHs, thermal fluctuations of the direction of the magnetization cause noise in addition to the familiar Johnson noise due to the conduction electrons' fluctuations. Both of these noise powers are derived from first principles and are then analyzed with the conclusion that the magnetization fluctuation noise will become a serious limitation in spin valves designed to operate at 100 gigabit/in.^{2}.

Colossal magneto-resistive (CMR) and electron spin tunneling heads, two potential candidates to eventually replace GMR spin valves, are considered in Chapter 20. The conclusion is that whereas the outlook for CMR is poor, the electron spin tunneling or magnetic tunneling junction (MTJ) heads appear both feasible and promising.

Finally, in Chapter 21, the extreme sensitivity of MRHs to accidental electrostatic discharges (ESDs) and the standard measures that may be taken to minimize the likelihood of ESD damage are reviewed.

An appendix which contains a listing of the defining equations and a table of conversion factors for cgs-emu and MKS-SI magnetic units is provided. Despite the fact that virtually all physicists and chemical engineers in the world are trained to use MKS-SI, the magnetic storage industries in the United States and Japan continue to use cgs-emu, and this convention is followed in this book. In my opinion, the preference for one convention over another is an unimportant matter of taste.

As in the first edition, the material in this book is presented in a basically nonmathematical manner. Thus there are no mathematical analyses or derivations. On the other hand, the results of such analyses and derivations are used extensively. My belief is that most readers are not interested in detailed mathematical formalism and usually find that copious mathematical detail acts to obscure the intuitive obviousness of the underlying physics. For example, the analysis of magnetization fluctuation noise in Chapter 19 is accomplished directly in only two algebraic operations. Rather than involving the reader in the esoterica of partition functions or the fluctuation-dissipation theorem, I treat the subject as an obvious consequence of the well-known equipartition theorem of thermodynamics. The lack of mathematics should not, however, lead the reader to suppose that this is an elementary exposition. On the contrary, the aim is to lead the reader, in a straightforward fashion, to a high level of scientific understanding of MR materials and heads.

Figures 11.5 and 11.6 and Tables 11.1 and 18.1 are reproduced with the permission of the IBM Corporation. The drawing on the front cover is by Ed Grochowski of IBM at Almaden and shows the cross sections of GMR spin valves for areal densities of 0.132, 0.354, 0.923, 1.45, 3.12, 5.7, 10.1, 17.1, >30, and >60 gigabit/in.^{2}, respectively.

Finally, my hope is that this book will prove enjoyable and fascinating to read. In the preface to the first edition I wrote: "Even after spending the last 35 years in magnetic recording research and theory, I find the seemingly unending inventiveness of my industrial colleagues both amazing and glorious." In the past six years, the inventiveness has continued unabated and the progress made in high-density MRHs is little short of staggering! I hope that this book helps others to share in this rich and fascinating field.

JOHN C. MALLINSON
BELMONT, CALIFORNIA

Acknowledgments

The author of this book acknowledges the help of his many friends in the magnetic recording industry. Karen Bryan produced all of the figures and Phebe Mallinson did the word processing.

CHAPTER 1

B, H, and M Fields

To understand magnetic recording and magneto-resistive heads (MRHs), we must first distinguish between the three fields B, H, and M. Each is a field which, at all points in three-dimensional space, defines the magnitude and direction of a vector quantity. The field B is called the magnetic flux density, H the magnetic field, and M the magnetization. All have properties similar to those of other, perhaps more familiar, fields such as the water flow in a river, the airflow over a wing, or the earth's gravitation.

Some arbitrariness exists concerning the order in which B, H, and M are introduced. Here the magnetic field is defined first, because it can be related to simple classical experiments. Magnetism and magnetization are treated next as logical extensions of the concept of a magnetic field. Finally, the magnetic flux density is discussed as the vector sum of the magnetic field and the magnetization.

Magnetic Field H and Magnetic Moment μ

When an electric current flows in a conductor, it produces a magnetic field in the surrounding space. Suppose that current I flows in a small length, dl, of the conductor. The magnetic field dH, measured at a distance r from the conductor, is orthogonal to both the current flow direction and the measuring distance vector. The magnitude of the magnetic field is given by the inverse square law and is

$$dH = 0.1 dl/r^2 \qquad (1.1)$$

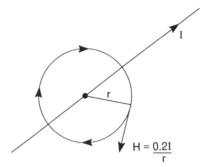

Fig. 1.1. The tangential field of a long, thin conductor.

where H is in oersteds, I is in amperes, and both l and r are in centimeters. All magnetic quantities in this book are given in the centimeter, gram, second, electromagnetic (cgs-emu) system of units. Conversion factors to the Système Internationale of metric units (SI) are given in the appendix.

Now consider a very long, straight conductor as shown in Fig. 1.1. By integrating Eq. (1.1), we can see that the magnetic field produced circles the conductor. The magnetic field is everywhere tangential, that is, orthogonal to both the current direction and the radius vector r. The direction of the magnetic field is given by the right-hand rule. Point the thumb of the right hand in the direction of the current flow, and the magnetic field direction is given by the way the curved fingers point. The magnitude of the field is given by

$$H = 0.2I/r, \tag{1.2}$$

where r is the radial distance in centimeters.

Now suppose that the conductor is coiled to form the solenoid shown in Fig. 1.2. The magnetic field inside a long solenoid is nearly uniform (parallel and of the same magnitude) and is

$$H = 0.4\pi NI/l, \tag{1.3}$$

where N is the number of turns and l is the solenoid length in centimeters. Note that the field does not depend on the cross-sectional area of a long solenoid. At this point, it is convenient to define another vector quantity, the magnetic moment,

$$\mu = \frac{1}{4\pi} \int H dV. \tag{1.4}$$

Electron Spin

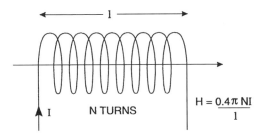

Fig. 1.2. The axial field of a long solenoid.

Here, μ is the magnetic moment in electromagnetic units (emu) and V is volume in cubic centimeters. For any solenoid, it can be shown that

$$\mu = 0.1 NIA, \qquad (1.5)$$

where A is the vector normal to the area of the solenoid cross section, whose magnitude is the solenoid cross-sectional area in square centimeters.

Electron Spin

Magnetization is a property which arises from the motions of electrons within atoms. The magnetization of *free space*, that is, space free of material bodies, is by definition zero.

In all atoms, electrons orbit a nucleus made up of protons and neutrons. The atomic number is the number of protons, which, for an electrically neutral (not ionized) atom, equals the number of electrons. The atomic weight is the number of protons plus the number of neutrons. Within an atom, the electron has two separate motions. First, the electron orbits the nucleus, much as the earth orbits the sun, at a radius of about 1 angstrom (10^{-8} cm). Second, the electron spins on its own axis, much as does the earth daily. These two motions are, of course, governed by the laws of quantum physics. All motions of an electron produce an electric current and, just as the motion of electrons around a solenoid produces a magnetic moment, the orbital motion gives rise to an orbital magnetic moment and the spinning motion causes an electron spin magnetic moment. Generally, the several electrons in an atom orbit in opposite directions, so that the total orbital moment is small. In the solid state, interactions with neighboring atoms "quench," that is, reduce further, the orbital moment. In the materials used in magnetic recording and magneto-resistive heads, the first transition group of elements, the contribution of the orbital magnetic moments to the magnetization is virtually negligible.

The magnetic moment of a spinning electron is called the *Bohr magneton* and is of magnitude

$$\mu_B = \frac{eh}{4\pi m} = 0.93 \times 10^{-20} \text{emu} \tag{1.6}$$

where e is the electron's charge in electromagnetic units (1.6×10^{-20}), h is Planck's constant (6.6×10^{-27}), and m is the electron's mass in grams (9×10^{-28}).

The spinning electron has a quantum spin number, $s = \frac{1}{2}$, and it can be oriented (in a weak magnetic reference field) in only $(2s + 1) = 2$ directions. For atoms in free space, the electrons' spin directions alternate, so that the total electron spin moment is no more than one Bohr magneton. Fortunately, however, nature has provided irregularities in the electron spin ordering. In transition group elements, outer electron orbits, or shells, begin to be occupied before the inner ones are completely filled. The result of this is that the inner partially filled shells can have large net electron spin moments and yet have the neighboring atom's interactions be partially screened off by the outer electrons.

Consider an atom of iron, the most common magnetic material, in free space, as depicted in Fig. 1.3. The atomic number of iron is 26, and there are, accordingly, 26 electrons. The atomic weight, 57, is the number of protons plus neutrons. The electron shells are shown numbered by two quantum numbers. The first, or principal, quantum number ($n = 1, 2, 3$, and 4) is related to the electron's energy. The orbital quantum number (s, p, and d) defines the orbital shape. Note that in shells $1s$, $2s$, $2p$, $3s$, $3p$, and $4s$, equal numbers of electron spins point up and down so that the total electron spin moment is zero; that is, the electron spins are "compensated."

In shell $3d$, however, according to Hund's rule, the first five spins are all oriented up with only one down and there is an uncompensated spin

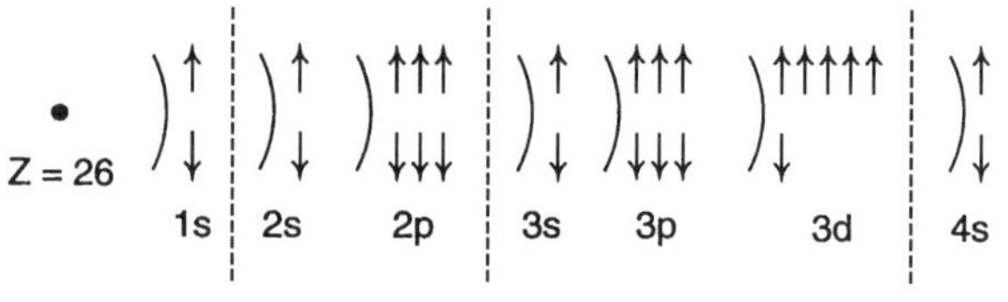

Fig. 1.3. The electron spin orientations of an iron atom in free space.

moment of 4 μ_B. Higher in the periodic table, the 3d shell eventually fills up with more down spins, resulting in 10 electrons with five up and five down and zero net electron spin moment. The first transition group elements (Cr, Mn, Fe, Co, and Ni), however, have 3d shells unfilled and have uncompensated electron spin magnetic moments. Nearly all practical interest in magnetism centers on the first and second transition groups of elements with uncompensated spins.

When iron atoms condense to form a solid-state metallic crystal, the electronic distribution, called the *density of states*, changes. Whereas the isolated atom has 3d; 5+, 1−, 4s; 1+, 1−, in the solid state the distribution becomes 3d; 4.8+, 2.6−, 4s, 0.3+, 0.3−. Note that the total number of electrons remains 8, but that the uncompensated spin moment is lowered to 2.2 μ_B. The screening of the neighboring atoms by the 4s electrons is imperfect. For all practical purposes, an iron atom in metallic alloys has 2.2 of μ_B magnetic moment.

Exchange Coupling

Now consider the magnetic behavior of iron atoms in an iron crystal. The crystal form, called the *habit*, is body-centered cubic with a cube-edge dimension of 2.86 Å, as determined by X-ray diffraction. The first question to ask concerns the relationship of the atomic moment of one iron atom to that of its neighbors. A quantum effect, called *exchange coupling*, forces all the iron atom's magnetic moments to point in nearly the same direction. Exchange coupling lowers the system's energy by aligning the uncompensated moments.

At absolute zero, the ordering is perfect, whereas at higher temperatures, thermal energy causes increasing disorder. At the Curie temperature, in iron 780°C, thermal energy equals the exchange energy and all long-range order breaks down. Thus, the spin moments point randomly in all directions. In this disordered state, the material is said to be a *paramagnet* (i.e., almost a magnet). Below the Curie temperature, the parallel alignment is called *ferromagnetism*.

Other orderings of the atomic moments are found in nature. The exchange coupling between atoms depends sensitively on the ratio of the interatomic distance to the atomic size as is indicated in Fig. 1.4. When the atoms are relatively closely spaced, as in the cases of Cr and Mn, the exchange coupling is negative and the adjacent spins are aligned in an antiparallel manner. Such materials are referred to as *antiferromagnets*

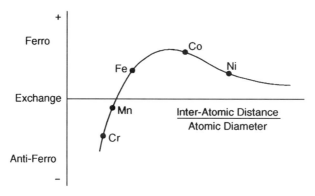

Fig. 1.4. The exchange coupling between magnetic atoms as a function of their spacing-to-size ratio.

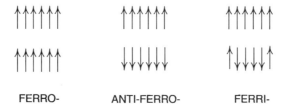

Fig. 1.5. Three common types of magnetic ordering.

and, of course, they have zero magnetic moment. Just as long-range order disappears in ferro- and ferrimagnets at the Curie temperature, in antiferromagnets, magnetic ordering vanishes at the Néel temperature.

For larger spacing ratios, the exchange coupling is positive and the spins are aligned parallel as in the ferromagnetic metals Fe, Co, and Ni. In yet other magnetic materials, for example, γ-Fe_2O_3 and the ferrites, an intermediate ordering called *ferrimagnetism* exists, where the number of spins in each direction is unequal. These three spin orderings are shown in Fig. 1.5.

Magneto-Crystalline Anisotropy K

Another question to ask about iron in the solid state concerns the orientation of the net magnetic moment with respect to the crystal axes. It turns out that the ferromagnetic-ordered atomic moments are aligned parallel to the body-centered cube edges. In iron, the cube edges are the easy, or lowest energy, directions, with the body diagonals being the hard, or highest energy, directions or axes of the magnetic moment. A measure of this energy difference is the magneto-crystalline anisotropy constant, K. It is the energy required, in ergs per cubic centimeter, to rotate the magnetic moments from the easy to the hard direction. Several different symmetries of magneto-crystalline anisotropy are found in nature. In magnetic recording technology, most interest centers on the cubic, as in iron and $\gamma\text{-}Fe_2O_3$, the hexagonal, as in cobalt alloys, and the uniaxial, as in iron–nickel (for example, permalloy) alloys.

Magnetization M

Now let our perspective expand to include a volume of iron that contains several million atoms. Just as previously our viewpoint enlarged from the electron spin level to the atomic level, now the focus is on millions of atoms. The magnetization vector is, by definition, the volume average of the vector sum of the atomic moments,

$$M = \frac{1}{V}\sum_1^N m, \qquad (1.7)$$

where V is the volume in cubic centimeters, m is the atomic moment in emu, and N is the number of atomic moments in the volume V. The units of magnetization are, therefore, magnetic moment per unit volume or electromagnetic units per cubic centimeter (emu/cm^3). In a large enough magnetic field, the magnetization of all parts of a magnetic material is parallel; at lower fields the magnetization may subdivide into domains. It is to be noted that within a single domain the magnetization is parallel everywhere and uniform and has a value called the saturation magnetization, M_s. The value of M_s depends on the temperature, being a maximum at absolute zero and vanishing at the Curie temperature, as the material becomes a paramagnet.

The 0°K value of M_s for a body-centered cubic iron crystal can be calculated easily. Each iron atom has 2.2 μ_B of magnetic moment; there are, on average, two iron atoms per unit cell; and the cell edges measure 2.86 Å. It follows that

$$M_s(T=0) = \frac{2.2 \cdot \mu_B \cdot 2}{(2.86 \times 10^{-8})^3} = 1700 \text{ emu/cm}^3 \qquad (1.8)$$

At room temperature, M_s is only slightly reduced by thermal energy, so that for all practical purposes, pure iron has the following properties: M_s = 1700 emu/cm^3, $4\pi M_s$ = 21,000 G, and σ_s = 216 emu/g, where σ_s is called the specific saturation magnetization.

The values of $4\pi M_s$ for some other materials of interest in magnetic recording are cobalt, 18,000 G; nickel, 6000 G; and permalloy (81 Ni/19 Fe), 10,000 G.

Magnetic Poles and Demagnetizing Fields

In general, when a magnetic material becomes magnetized by the application of a magnetic field, it reacts by generating, within its volume, an opposing field that resists further increases in the magnetization. This opposing field is called the *demagnetization field* because it tends to reduce or decrease the magnetization. To compute the demagnetization fields, first the magnetization at all points must be known. Then, at all points within the sample, the magnetic pole density is calculated,

$$\rho = -\nabla \cdot M = -\left(\frac{dM_x}{dx} + \frac{dM_y}{dy} + \frac{dM_z}{dz}\right), \tag{1.9}$$

where ρ is the pole density (emu/cm^4), and M_x, M_y, and M_z are the orthogonal components of the magnetization vector. The convention for magnetic poles is that when the magnetization decreases, the poles produced are north or positive. It is an unfortunate historical accident that the earth's geographic north pole has south, or negative, magnetic polarity.

Magnetic poles are of extreme importance because they also generate magnetic fields, H. The only two sources of magnetic fields are *real* electric currents and magnetic poles. The adjective *real* is used to distinguish real currents flowing in wires, which may be measured with ammeters, from the circulating currents flowing in atoms due to their orbiting and spinning electrons which cause magnetization to exist. The magnetic fields caused by magnetic poles can be computed using the inverse square law. The field points radially out from the positive or north pole and has the magnitude

$$dH = \rho dV/r^2, \tag{1.10}$$

where H is the magnetic field in oersteds, ρ is the pole density, V is the volume (cm^3), and r is the radial distance (cm). Note that magnetic poles are analogous to electric poles.

Magnetic Poles and Demagnetizing Fields

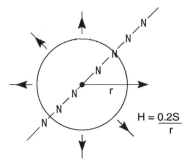

Fig. 1.6. The radial field from a long line of poles.

Most field computations in magnetic recording are two dimensional. This is because one dimension, the track width, is very large compared with the other two. Two-dimensional magnetic fields have many simplifying properties, and one of the most important of these is shown as Fig. 1.6. The magnetic field from a long straight line of poles points radially out and has magnitude

$$H = 0.2s/r, \qquad (1.11)$$

where H is the magnetic field in oersteds, s is the pole strength per unit length (emu/cm^2), and r is the radial distance in centimeters. The crucial thing to notice is that Eqs. (1.2) and (1.11) have the same form. Apart from scaling factors, a change from electric current to magnetic poles causes a 90° rotation of the magnetic field at every point in a two dimensional space. The magnetic fields from currents and poles are orthogonal.

In general, the fields generated by a magnetic body are very complicated and force the magnetization to be nonuniform. For one class of geometrical shapes, however, it is known that the demagnetizing field and magnetization can be uniform. When any ellipsoid is uniformly magnetized, the demagnetizing field is also uniform. The demagnetizing field can be written

$$H_d = -NM, \qquad (1.12)$$

where H_d is the vector demagnetizing field, N is the demagnetization tensor, and M is the vector magnetization. For all ellipsoids, the demagnetization tensor is the same at all points within a given body. Note that since "tensors turn vectors," the demagnetizing field need not be exactly antiparallel to the magnetization. The special class of ellipsoids of revolution range from infinite flat plates through oblate spheroids to spheres, through prolate spheroids to infinite cylinders. They are formed by rotating an ellipsoid about either its major or minor axis. The demagnetizing tensors for three cases are shown below:

xx xy xz	0 0 0	$4\pi/3$ 0 0	2π 0 0
yx yy yz	0 0 0	0 $4\pi/3$ 0	0 2π 0
zx zy zz	0 0 4π	0 0 $4\pi/3$	0 0 0
Tensor	Flat Plate	Sphere	Long Cylinder

Thus, the flat plate has no demagnetization within its x–y plane but suffers a 4π demagnetizing factor on magnetization components out of the plane. A sphere suffers a $4\pi/3$ factor in all directions. A long cylinder has no demagnetization along its axis, but suffers 2π in the x and y directions of its cross sections. Note that these tensors are all diagonal, because the axis of rotation coincides with the z direction, and that the diagonal terms always sum to 4π. This is because 4π steradians of solid angle fill three dimensional space. In cgs-emu, 4π field lines emanate from a unit magnetic pole. In other systems of units (e.g., MKS-SI) the 4π appears in other places, but its appearance cannot, of course, be suppressed entirely, as is shown in the appendix, Table A.2.

Consider an ellipsoidal sample of magnetic material within a long solenoid that produces an axial field, H_s, as shown in Fig. 1.7. Suppose that the solenoid field is strong enough to saturate the magnetization. The magnetization of the ellipsoid is uniform, and thus there are no magnetic poles within the volume. Poles form, however, on the surfaces as shown by the letters N and S in the figure. The surface pole density is n·M (emu/cm^3), where n is the outward pointing, normal, unit vector on the surface. These surface poles produce a demagnetizing field, H_d, which is exactly antiparallel

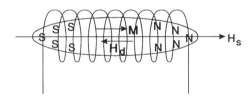

Fig. 1.7. An ellipsoid of revolution in a solenoid, showing the induced surface magnetic poles and the demagnetizing field.

to both the magnetization and the solenoid field. If the ellipsoid axis had not been parallel to the solenoid axis, these exact alignments would not have occurred. It is clear now that the total magnetic field, H_t, within the sample is

$$H_t = H_s - H_d. \tag{1.13}$$

The effect of the demagnetizing process is to reduce the field inside the sample in a manner which is exactly analogous to that of the negative feedback system.

Flux Density B and Flux ϕ

Now that both the magnetic field H and the magnetization M have been defined and discussed, the flux density B can be defined as

$$B = H + 4\pi M, \tag{1.14}$$

where B is the flux density in gauss, H is the (total) field in oersteds, and M is the magnetization in electromagnetic units. In this equation, B, H, and M are all field vector quantities and addition is performed vectorially.

The M field, the H field, and the B field of a uniformly magnetized bar magnet are shown in Fig. 1.8. In all the cases, the scheme adopted for showing, or plotting, the field in the plane of the paper is the same. At all points, the lines and arrows show the direction of the particular field quantity. The spacing between the lines is inversely proportional to the field magnitude. The closer the lines, the higher the field strength. In this scheme,

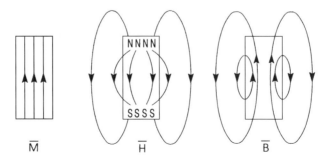

Fig. 1.8. The magnetization, magnetic field, and flux density fields of a bar magnet.

which is identical to that used to depict, for example, airflows, the "flow" or "flux" of the field quantity between all adjacent pairs of lines is equal. In dynamic fields, such as airflows, the lines are called *streamlines*. The *B*, *H*, and *M* fields are static, but the same nomenclature persists. The streamlines are also called *lines of force*, and between them, for the *B* field, flows the magnetic flux ϕ where

$$\phi = \int B \cdot dA. \qquad (1.15)$$

The *M* field pattern of Fig. 1.8 shows parallel lines within the uniformly magnetized magnet only. The magnetization outside the magnet is zero. The *H* field drawing shows the magnetic poles caused by the changing magnetization on the ends of the magnet. Where *M* is decreasing, north, or positive, poles arise as at the top of the magnet. The convention is that *H* lines of force emanate from the north, or positive, poles. Note that inside the body of the magnet, the magnetic field looks very similar to the electric field between two electrically charged capacitor plates and is not uniform. This is because the bar magnet's shape is not ellipsoidal.

The magnetic field inside the magnet generally opposes the magnetization and is, appropriately enough, called the *demagnetizing field*. The magnetic field outside the magnet, which arises from the very same magnetic poles causing the demagnetizing field inside, is often called the *fringing* or *stray field*. Because both have the same origin, it is clear that large external fringing fields imply high internal demagnetizing fields.

The *B* field plot is, of course, identical to that of the *H* field for all the points outside the magnet. This is because in free space, $B = H$, there being no magnetization. It follows that it is immaterial whether one speaks of the field in the gap of an electromagnet or a magnetic writing head as being *B* in gauss or *H* in oersteds.

Inside the magnet, the vector addition of *H* and $4\pi M$ produces the converging flow shown in Fig. 1.8. Note, particularly, that unlike the *H* and *M* fields, the *B* field flows continuously and has no sources or sinks, that is $\nabla \cdot B = 0$. This is a consequence of the fact that isolated magnetic monopoles do not exist. North and south poles coexist in equal numbers as dipoles.

The slope of a ***B*** versus ***H*** curve, *dB/dH*, is called the permeability, μ, so that in the linear case, $B = \mu H$.

Further Reading

Additional reading material is listed here that will prove helpful to readers who seek more detailed information.

Bozorth, Richard M. (1951), *Ferromagnetism,* Van Nostrand-Reinhold, Princeton, New Jersey. (Available from University Microfilms.)

Jiles, David (1991), *Introduction to Magnetism and Magnetic Materials,* Chapman and Hall, London.

Smit, J., and Wijn, H. P. J. (1959), *Ferrites,* Wiley, New York.

CHAPTER

2

The Writing Process

In this chapter, the fundamentals of the writing process are reviewed. This review is of great importance because some aspects of the design of MRHs depend on the written magnetization transition. In order to obtain the maximum output voltage from an MRH, it *must match properly* the written magnetization transition in the recording medium. The topic of this proper matching is pursued extensively in Chapters 17 and 18.

Most applications of MRHs are in digital binary recording. Accordingly, this chapter deals only with the writing process in digital recording.

Write-Head Field

The stylized writing head, shown in Fig. 2.1, consists of three parts: the core, the coil, and the gap.

In ferrite heads the core is usually made of NiZn or MnZn ferrite. These materials are electrical insulators which can be operated at high frequencies (>50 MHz) without requiring thin laminations. In thin-film heads, the core structure is usually called the yoke. They are made of thin layers of permalloy (81Ni/19Fe) or AlFeSil (an aluminum, iron, and silicon alloy) in, typically, 2- to 4-μm thicknesses. These permit operation at frequencies as high as 200 MHz without appreciable eddy current losses. Both permalloy and AlFeSil have saturation flux densities, $10,000 < B_s < 12,000$ gauss.

Write-Head Field

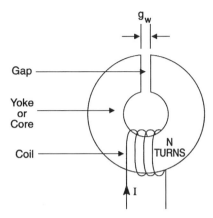

Fig. 2.1. An idealized writing head showing the core or yoke, the coil, and the gap.

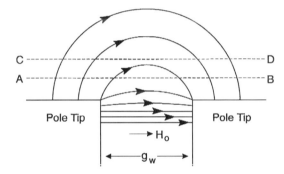

Fig. 2.2. The fringing field around the top of the gap of a write head.

Usually a ten turn coil carries the writing current, I_w, which is typically of magnitude 10 to 20 mA peak. The writing current is toggled from one polarity to the other in order to write digital transitions of the remanent magnetization in the recording medium. Making clear the nature of these transitions is the principal purpose of this chapter. The inductance of a head depends on the square of the number of turns in the coil. Because high-voltage write amplifiers are required in order to drive high-inductance heads, the normal design trend is to use a small number of turns and high current rather than vice versa.

The gap, g_w, permits the magnetic flux circulating in the core to fringe out and intercept the recording medium as shown in Fig. 2.2. When the head-to-medium spacing and medium thickness are held constant, changing the write gap length g_w has little effect on the written transition. This is because when the coil current I_w is changed simultaneously, so that the same value of the deep gap field, H_0, is maintained, the geometry of the active writing zone over the trailing pole gap edge changes little. In practice, the gap length is determined by a trade-off between better overwriting (larger gap) and lower side-writing (smaller gap).

The deep gap field H_0 in oersteds is

$$H_0 = \frac{0.4\pi N I \eta}{g_w}, \tag{2.1}$$

where N is the number of turns in the coil, I is the writing current in amperes, g_w is the gap length in centimeters, and η is the writing head efficiency, which is the fraction of the amp-turns in the coil that appears across the gap. η is usually >0.8 in well-designed heads.

In high-density recording, the deep gap field required is

$$H_0 = 3H_c, \tag{2.2}$$

where H_c is the coercivity of the recording medium as shown in Fig. 2.3. To prevent appreciable magnetic saturation in the pole tips, it is necessary that

$$H_0 \leq 0.6 B_s, \tag{2.3}$$

where B_s is the saturation flux density of the pole or yoke material. High B_s materials, such as 45Ni/55Fe (16,000 G) and Fe_7 AlN (19,000 G), permit higher H_c media to be used.

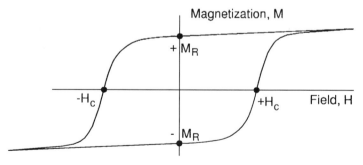

Fig. 2.3. The magnetization versus field hysteresis loop of recording medium.

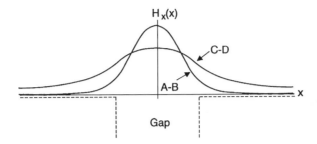

Fig. 2.4. Plots of the horizontal component of the fringing field versus distance along the top of the head.

As long as pole tip saturation is avoided, the horizontal component, H_x, of the fringing field above the gap at a point P is

$$H_x = \frac{H_0}{\pi}\tan^{-1}\left(\frac{yg_w}{x^2 + y^2 - g_w^2/4}\right), \qquad (2.4)$$

where x and y are the horizontal and vertical coordinates of the point P. The origin of the coordinates is at the top and on the centerline of the gap. In Fig. 2.4 the horizontal component H_x is plotted versus horizontal distance x for two cases where the point P travels across the top of the write head along trajectories A–B and C–D. Note that the trajectory closer to the head (A–B) has both a higher maximum field and higher field gradients, dH_x/dx.

Written Magnetization Transition $M(x)$

When the write current is held constant, the magnetization written in the recording medium is at one of the remanent levels, M_R, shown in Fig. 2.3. When the write current is suddenly changed from one polarity to the other, the written magnetization undergoes a transition from one polarity of remanent magnetization to the other.

Here it is assumed that the transition has the form shown in Fig. 2.5,

$$M(x) = \frac{2}{\pi}M_R\tan^{-1}\left(\frac{x}{f}\right), \qquad (2.5)$$

where f is the so-called "transition slope" parameter. Note that, as shown in Fig. 2.5, the maximum slope, which occurs at $M = 0$, is $2M_R/\pi f$ and

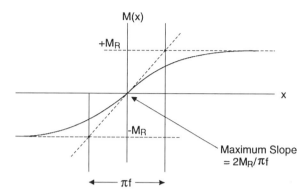

Fig. 2.5. The arctangent magnetization transition showing the maximum slope and the approximate width.

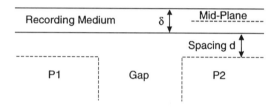

Fig. 2.6. The geometry of the writing process, showing the head-to-medium spacing and the midplane of the thin recording medium.

that a straight-line approximation of the transition has a width of πf. As f is reduced, the transition becomes steeper and more closely approaches the binary ideal of a step function, when $f = 0$.

The geometry of the write process is shown in Fig. 2.6. The write head-to-medium spacing is d and the recording medium thickness is δ. The gap length between the two poles P1 and P2 is not critical, but is assumed small in the discussion below.

The write current is adjusted so that the horizontal component of the fringing field on the midplane of the recording medium meets a specific criterion. This criterion, shown in Fig. 2.7, is that the horizontal position, where the field magnitude is equal to H_c, must coincide with the position where the head field gradient, dH_x/dx, is greatest. Meeting this criterion sets both the magnitude of the write current I_w and the horizontal position of the center of the magnetization transition $[x = (d + \delta/2)/\sqrt{3}]$.

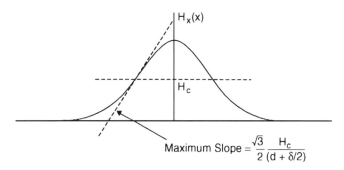

Fig. 2.7. The optimized write-head field versus distance showing that the maximum slope occurs at the coercivity.

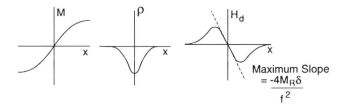

Fig. 2.8. Plots of the magnetization, the pole density, and the demagnetizing field of a written transition.

The maximum head field gradient is

$$\left(\frac{dH_x}{dx}\right)_{max} = \frac{\sqrt{3}}{2}\frac{H_c}{(d+\delta/2)}. \tag{2.6}$$

The written magnetization transition, M versus x, is shown again in Fig. 2.8. Also shown is the magnetic pole density, ρ versus x, and the demagnetizing field on the centerline of the medium, H_d versus x. Note that because the magnetization increases through the transition, the pole density is negative and has south polarity. The maximum slope or gradient of the demagnetizing field is

$$\left(\frac{dH_d}{dx}\right)_{max} = \frac{-4M_R\delta}{f^2}. \tag{2.7}$$

The maximum gradient occurs at the center ($M = 0$) of the transition. The smaller the value of the slope parameter f, the higher the magnitude of the demagnetizing field and its gradient.

In the Williams and Comstock model of the digital write process, the "slope" equation is used, where

$$\frac{dM}{dx} = \frac{dM}{dH} \cdot \frac{d(H_{\text{head}} + H_{\text{demag}})}{dx}. \tag{2.8}$$

For a square loop recording medium, *dM/dH* is very high, and a convenient approximation is to set the maximum head field gradient equal to the maximum demagnetizing field gradient. Upon setting Eqs. (2.6) and (2.7) equal, the result is

$$f = 2\left[\frac{2}{\sqrt{3}}\frac{M_R}{H_c}\delta(d+\delta/2)\right]^{1/2}. \tag{2.9}$$

The writing problem is now completely solved because f is but the single parameter required to define fully the magnetization transition of Eq. (2.5). Note that possible ways to reduce the transition width, by reducing f, are to use higher coercivities, lower remanences, smaller flying heights, and thinner media. With the exception of lowering the remanence, all have been exploited in the past. When inductive reading heads are used, reducing the remanent magnetization is not an acceptable strategy, however, because it always reduces the signal and signal-to-noise ratio of the recorder. With MRHs, however, lower remanence disks may be used as is explained in Chapters 17 and 18.

Equation (2.9) for the transition slope parameter f is used in both the simplified design of the shielded anisotropic magneto-resistive heads and the shielded giant MR spin valve heads treated in Chapters 17 and 18, respectively.

Further Reading

Additional reading material is listed here that will prove helpful to readers who seek more detailed information.

Williams, M. L., and Comstock, R. L. (1971), "An Analytical Model of the Write Process in Digital Magnetic Recording," in 17*th Annual Conf. Proc.*, pp. 738–742, American Institute of Physics (also in R. M. White; see Bibliography).

CHAPTER 3

The Reading Process

The reading process in inductive reading heads is reviewed in this chapter for several reasons. First, it is necessary to know the fringing fields surrounding the written magnetization in the recording medium in order to understand the behavior of unshielded MRHs. Second, when the magnetic flux flowing around a normal inductive reading head is known, then it is possible to deduce how shielded MRHs work in an extremely elegant way. Third, a knowledge of the output voltage spectrum and the isolated magnetization transition pulse shape of an inductive head facilitates direct comparisons with the equivalent properties of MRHs.

Recording Medium Fringing Fields

The fringing field or flux of a recorded medium is shown in Fig. 3.1. The written magnetization waveform is indicated as the dashed line. Where this magnetization decreases, north poles occur, and vice versa for south poles.

The magnetic field and flux fringes equally above and below the medium, flowing from the north to the south poles. Suppose the written magnetization waveform is

$$M_x(x) = M_R \sin kx, \qquad (3.1)$$

where M_R is the maximum amplitude of the written magnetization and k is the wavenumber ($=2\pi/\lambda$), where λ is the sinusoidal wavelength.

Fig. 3.1. The fringing field above and below a recording medium with a written magnetization pattern (dashed line).

The horizontal and vertical components of the fringing field at a point (x, y) below the medium are

$$H_x(x, y) = -2\pi M_R(1 - e^{-k\delta})e^{-ky}\sin kx \qquad (3.2A)$$

and

$$H_y(x, y) = 2\pi M_R(1 - e^{-k\delta})e^{-ky}\cos kx. \qquad (3.2B)$$

These expressions for H_x and H_y will be used in Chapter 8 for the analysis of Hunt's unshielded MRHs.

Read-Head Flux

When a reading head made with relatively massive pole pieces of highly permeable material is placed close to the recording medium, the field and flux flow patterns are changed profoundly as shown in Fig. 3.2. Relatively massive here means large compared to the recorded wavelength λ. With the reading head in place, almost all the fringing flux flows into the pole pieces, P1 and P2, because of their "keepering" action.

The reading head has the same structure as the writing head discussed in the last chapter. Because the pole pieces have a high magnetic permeability

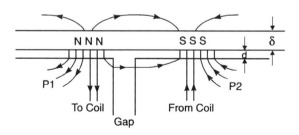

Fig. 3.2. The flux flow around the top of the gap in a highly permeable reading head.

($\mu = dB/dH$), most of the fringing flux flows deep in the head passing through the coil. Very little flux flows through the nonmagnetic gap. The ratio of the flux passing through the coil to the flux entering the top surface of the head is called the *read-head efficiency*. In inductive heads, the efficiency is normally in excess of 80%.

For the sinusoidal recording, the flux ϕ threading the coil of a 100% efficient reading head is

$$\phi(x) = -4\pi M_R W \frac{(1-e^{-k\delta})e^{-kd}}{k} \frac{\sin kg/2}{kg/2} \sin kx, \qquad (3.3)$$

where d is the head-to-medium spacing, g is the gap length, and w is the track width.

Output Voltage

The time rate of change of the flux linkages, $N\phi$, in a head coil with N turns is proportional to the read-head's output voltage, E. Thus,

$$E = -10^{-8}\frac{d(N\phi)}{dt} = -10^{-8}N\frac{d\phi}{dt} = -10^{-8}NV\frac{d\phi}{dx} \quad [\text{volts}], \quad (3.4)$$

where V is the head-to-medium relative velocity. On putting Eq. (3.3) into Eq. (3.4), the well-known result is

$$E(x) = 10^{-8}NVW4\pi M_R(1-e^{-k\delta})e^{-kd}\frac{\sin kg/2}{kg/2}\cos kx. \qquad (3.5)$$

Note that the output voltage is proportional to the number of coil turns N, the head-to-medium velocity V, and the written remanence M_R.

The term in parentheses in Eq. (3.5) is called the *thickness loss*, and it shows that the read head is unable to sense magnetization patterns written deep into the medium. The exponential term e^{-kd} is called the *spacing loss*, and it is often quoted as $-55d/\lambda$ dB.

The factor $\sin kg/2/(kg/2)$ is called the *gap loss*. At the first gap null, at wavelength $\lambda = g$, the gap loss term is equal to zero. The fact that the output voltage waveform is a cosine when a sine wave is written shows that the phase of the output signal is lagging 90° behind the written magnetization.

Output Spectrum

The peak magnitude of the output voltage as a function of frequency is called the *spectrum*. The temporal frequency $f = V/\lambda$ and the angular frequency $\omega = 2\pi f$. The spatial frequency or wavenumber $k = 2\pi/\lambda = \omega/V$ so that $\omega = Vk$.

The spatial frequency spectrum corresponding to Eq. (3.5) is just $A(k) = E(x)/\cos kx$ and it is shown in Fig. 3.3. Note that it has zeros at both dc and the first gap null.

Digital Output Pulse

When the reading head passes over a written magnetization transition of the type discussed in Chapter 2, the coil flux and output voltage are as shown in Fig. 3.4.

The Fourier transform of the magnetization transition

$$M(x) = \frac{2}{\pi} M_R \tan^{-1}\left(\frac{x}{f}\right) \tag{3.6A}$$

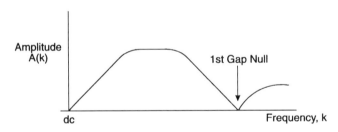

Fig. 3.3. The amplitude of the output voltage spectrum versus wavenumber showing dc and gap nulls.

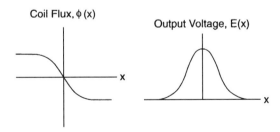

Fig. 3.4. The flux flowing in and output voltage of a head reading a magnetization transition.

is

$$M(k) = 2M_R e^{-kf}/jk. \quad (3.6B)$$

Here f is the transition slope parameter and $j = \sqrt{-1}$. The Fourier transform, $M(k)$, is simply the spatial frequency domain representation of the x domain transition $M(x)$.

Because the inductive head reading process is strictly linear, the output signal spectrum when reading the magnetization transition can be written immediately:

$$0(k) = M(k)E(k), \quad (3.7)$$

where $E(k) = jA(k)$ is the read-head output voltage spectrum. The digital output pulse, $0(x)$, is the inverse Fourier transform of Eq. (3.7) and is shown in Fig. 3.5. Note that the pulse is symmetrical for $\pm x$. The negative "undershoots" shown are a direct consequence of the fact that a reading head, regardless of whether it is inductive or magneto-resistive, cannot have a dc response.

When the read-head gap length g is small, the output pulse is

$$0(x) = 10^{-8} NVW 4\pi M_R \log_e \frac{(d+f+\delta)^2 + x^2}{(d+f)^2 + x^2}. \quad (3.8)$$

The 50% pulse width, PW$_{50}$, measured at 50% of the peak amplitude, is

$$PW_{50} = 2[(d+f+\delta)(d+f) + g^2/4]^{1/2}. \quad (3.9)$$

Equations (3.8) and (3.9) are used widely later in this book.

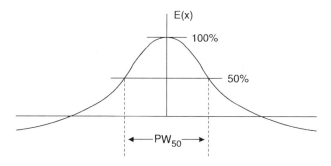

Fig. 3.5. The isolated transition output pulse showing the inevitable negative undershoots.

As the alternating polarity transitions are packed closer, the output voltage falls. A convenient approximation is that $PW_{50} \cdot D_{50} \approx 1.45$, where D_{50} is the linear transition density with peak voltage exactly 50% that of Eq. (3.8).

Further Reading

Additional reading material is listed here that will prove helpful to readers who seek more detailed information.

Wallace, R. L. (1951), "The Reproduction of Magnetically Recorded Signals," *Bell Syst. Tech. J.* (also in R. M. White; see Bibliography).

CHAPTER

4

The Anisotropic Magneto-Resistive Effect

The Basic Effect

In 1857, William Thomson, who later became Lord Kelvin, discovered that the electrical resistance of an iron bar changed as its magnetization was altered. This phenomenon is now called the *anisotropic magnetoresistive* (AMR) effect, and it remained a mere academic curiosity until Bob Hunt invented the anisotropic magneto-resistive head (AMRH).

In a thin film of magnetic material, the magnetization can be single domain and not be subdivided by domain walls into the multidomain state. Moreover, in a thin film, demagnetizing fields force the magnetization direction to remain virtually parallel to the plane of the film.

In the thin-film single-domain state, the AMR effect can be described very simply. With reference to Fig. 4.1, when the angle between the magnetization, M, and the electrical measuring or sensing current, I, is θ, the electrical resistance, R, can be written

$$R = R_0 + \Delta R \cos^2 \theta, \tag{4.1}$$

where R_0 is the fixed part and ΔR is the maximum value of the variable part of the resistance, respectively.

In Fig. 4.1, note that the thickness of the thin film is T, the film depth is D, and the film width is W. This labeling of the MR sensor or MR element

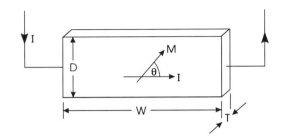

Fig. 4.1. The AMR effect in a single-domain thin-film strip.

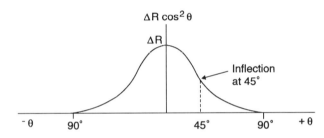

Fig. 4.2. The magneto-resistive change in resistance versus magnetization angle.

(MRE) is used consistently throughout this book. The width direction is usually both the electrical current direction and the recorded track-width direction.

A plot of the variable part of the resistance of a material like permalloy versus the magnetization angle is shown in Fig. 4.2. Because the maximum value of the variable part of the resistance occurs at $\theta = 0$, it is seen that, perhaps counter to intuition, the resistance is maximum when the magnetization and current are parallel. Note that the resistance change versus angle characteristic has an inflection point at $\theta = 45°$ where the characteristic is, for small angular excursions, almost linear.

Magneto-Resistive Sensors or Elements

The fundamental idea in all AMRHs is that the magnetic field produced by the written or recorded magnetization pattern in the tape or disk rotates the MRE magnetization angle. When the magnetization angle in the MRE is properly biased with a vertical bias field, produced by the means discussed in Chapter 6, small changes in resistance, δR, are almost directly proportional to small amplitude fields from the recording medium.

Magneto-Resistive Sensors or Elements

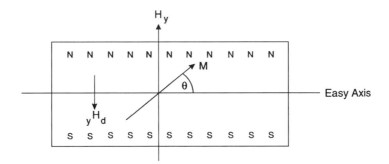

Fig. 4.3. An MRE showing the easy axis, the disk field, the magnetic poles at the edges, and the vertical demagnetizing field.

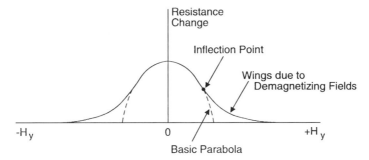

Fig. 4.4. The magneto-resistive change in resistance versus disk field, showing the effects of the nonuniform demagnetizing field.

In all MRHs, the output signal voltage is, by Ohm's law, equal to the resistance change times the measuring current: $\delta V = I \delta R$.

When the magnetization angle θ in the MRE increases, magnetic poles are generated in the top and bottom regions of the sensor, as shown in Fig. 4.3. These poles, of north polarity at the top and south at the bottom, generate an internal field, usually called the *demagnetizing field*, $_y H_d$. If the vertical magnetic field, H_y, which is the sum of the vertical bias field and the tape or disk field, is acting up, the demagnetizing field induced by the magnetization angle θ is down as shown in Fig. 4.3. Whereas the vertical field acts to increase θ, the demagnetizing field acts to reduce θ. The value of the demagnetizing field, averaged over the element depth D, is equal to $\frac{4\pi M_s \sin\theta\, T}{D}$. Thus, the thicker and shallower the MRE, the greater the demagnetizing fields for a given magnetization angle θ.

Figure 4.4 shows the change in resistance versus vertical field, H_y. If the demagnetizing fields are negligible (thin and deep element), the magnetization angle is

$$\theta = \sin^{-1}(H_y/H_k) \qquad (4.2)$$

and the resistance is

$$R = R_0 + \Delta R[1 - (H_y/H_k)^2]. \qquad (4.3)$$

Here H_k is the fictitious "anisotropy field" discussed later in this chapter. The variable part of the resistance is parabolic, shown as a dashed line in Fig. 4.4, and has no inflection point. The effect of taking the nonuniform demagnetizing fields into account is the appearance of the wings shown as solid lines. More vertical field is required to achieve a given magnetization angle. Note that an inflection point once more exists. At the inflection point, the magnetization angle might typically vary from almost 0° at the top and bottom of the element to perhaps 60° at the middepth, but the average angle is 45°.

Provided the vertical field is not sufficient to saturate the MRE, that is, $M_y < M_{sat}$ at the middepth $y = D/2$, an exact analytical solution for the magnetization angle θ is known for the usual case where the anisotropy field H_k is negligible compared with the demagnetizing field $_yH_d$. The magnetization angle as a function of element depth y is

$$\theta(y) = \tan^{-1}\left\{\frac{H_{bias}}{2\pi M_s T}\left[\left(\frac{D}{2}\right)^2 - y^2\right]^{\frac{1}{2}}\right\}. \qquad (4.4)$$

For many purposes, it is both convenient and sufficient to approximate the change in resistance versus vertical field characteristic with the straight heavy solid line shown in Fig. 4.5. The slope of this approximated

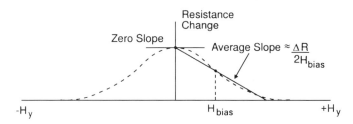

Fig. 4.5. The MR resistance change versus the disk field characteristic, showing the average slope and the bias field.

Magneto-Resistive Sensors or Elements

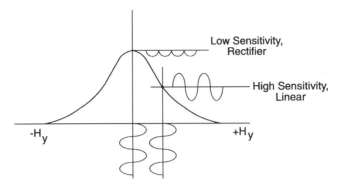

Fig. 4.6. A schematic representation of the MR reading of disk fields with and without a bias field.

characteristic is equal to $-\frac{1}{2}\Delta R/_y H_{bias}$ and it represents the sensitivity of the MRE when vertical bias is used. This approximation was made in the first analysis of Hunt's unshielded AMRHs, the results of which are discussed in Chapter 8.

When no vertical bias is used, the sensitivity, or slope, of the resistance change versus field characteristics is, of course, zero.

Figure 4.6 depicts the small resistance changes δR in response to oscillating polarity small fields from a written magnetization pattern in a tape or disk. When the proper vertical bias field is used, the output voltage, $I\delta R$, is both large and almost linear. Typically, deviations from linearity cause about −20 dB of even harmonic distortion, which, though satisfactory for a binary or digital data channel, is not sufficiently linear for an analog signal channel. On the other hand, if vertical biasing is not used, the response is of low sensitivity and is highly nonlinear, behaving as a full-wave rectifier.

It is important to realize that even when the optimum vertical bias is used, the output of an MRE is inherently nonlinear even for small tape or disk fields. In this respect an AMRH differs greatly from the usual inductive reading head. Accordingly, simple AMRHs always show pulse amplitude asymmetry and even harmonic distortion with the positive and negative output pulses being of unequal amplitude. Furthermore, the $PW_{50}s$ of the opposite polarity output pulses must be unequal. Several schemes for minimizing this inherent asymmetry are discussed in Chapters 6 and 15 of this book.

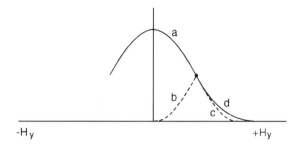

Fig. 4.7. The MR resistance change versus field characteristic, showing the lack of symmetry about the bias point.

The precise origins of both the odd and even harmonic distortion can be understood with the help of Fig. 4.7. Suppose the upper portion of the resistance change characteristic labeled (a) is folded or reflected about an imaginary horizontal plane to form the dashed line marked (b). Let the reflected part (b) be folded again, about an imaginary vertical plane, to form the second dashed line (c).

Now if the twice-folded dashed line (c) happened to coincide with the lower portion of the characteristic labeled (d), odd harmonic distortion only would occur in the MRH response. Odd harmonic distortion only causes the output pulses to be clipped or limited and does not cause pulse amplitude asymmetry.

In reality, however, dashed section (c) does not match the lower part of the characteristic (d). The mismatch causes the generation of even harmonics and the output pulses to be of unequal amplitudes.

The mismatch shown in Fig. 4.7 causes a negative H_y pulse to have a larger amplitude than that of a positive H_y. This is because part (a) has a greater negative slope than does part (d).

All present AMRHs use either vertical bias or other means to achieve quasi-linear operation. Most of the schemes which have been proposed to achieve vertical biasing are discussed in Chapter 6.

Just as the finite depth D of the MRE causes the vertical demagnetizing fields $_yH_d$, as already discussed, the finite width W causes a horizontal demagnetizing field, $_xH_d$. This is shown in Fig. 4.8.

Both the vertical and horizontal demagnetizing fields depend on the MRE dimensions and magnetization angle θ, and they vary spatially. Plots showing this variation and the average value of the vertical and horizontal demagnetizing fields are shown schematically in Figs. 4.9 and 4.10, respectively.

Magneto-Resistive Sensors or Elements

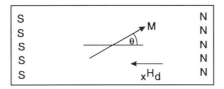

Fig. 4.8. An MRE showing the end-zone poles and the horizontal demagnetizing field.

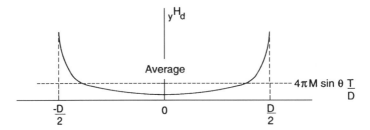

Fig. 4.9. A plot of the vertical demagnetizing field versus depth, showing the average value.

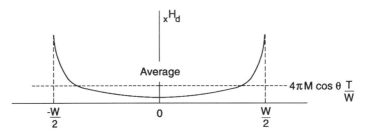

Fig. 4.10. A plot of the horizontal demagnetizing field versus width, showing the average value.

To keep the single-domain magnetization in the MRE stable and free from hysteresis, it is usually necessary to provide a horizontal bias field in addition to the vertical bias field already discussed. The existence of hysteresis causes the resistance change for increasing magnitude vertical fields to be different from that for decreasing vertical fields. The presence of hysteresis indicates that, in some part of the MRE, usually at the ends, there are domain walls which have moved discontinuously from one pinning site to another. This discontinuous motion causes jumps in the MRE magnetization angle, resistance, and output voltage. These jumps are usually termed Barkhausen noise. It is essential that hysteresis and Barkhausen jumps be eliminated, and several of the means proposed for achieving this are discussed in Chapter 7. Exactly the same considerations apply to the free or sensor layer in GMR spin valve heads.

Magneto-Resistive Coefficient $\Delta\rho/\rho_0$

The electrical resistance, R, of the MRE, in terms of the resistivity, ρ, of the material, is

$$R = \rho W/TD. \qquad (4.5)$$

In AMR materials

$$\rho = \rho_0 + \Delta\rho \cos^2\theta, \qquad (4.6)$$

where ρ_0 is the fixed part and $\Delta\rho$ the maximum value of the variable part of the resistivity.

The magneto-resistive coefficient is defined to be $\Delta\rho/\rho_0$ and it is the maximum fractional change in the resistivity. For many purposes, this coefficient serves as a figure of merit of the MR material. Note that $\Delta\rho/\rho_0$ equals $\Delta R/R_0$.

Permalloy is a binary alloy of nickel and iron of nominal composition 81Ni/19Fe. The magneto-resistive coefficient of permalloy is approximately 4% in thick (>1000-Å) films. In thinner films, the coefficient decreases due to the increasing importance of surface scattering of the conduction electrons, which mainly increases the fixed part of the resistivity with $\Delta\rho$ remaining constant. For the 100 to 200-Å-thick films frequently used today, the MR coefficient is only about 2%.

It is known that the AMR originates in the anisotropic scattering of d shell conduction electrons in the exchange split d bands. As was discussed in Chapter 1, the d shells contain the electrons which are responsible for ferromagnetism and ferrimagnetism. Exchange split merely means that the energy of the up and down spin conduction electrons differs by the quantum-mechanical exchange energy discussed in Chapter 1.

Seemingly convincing analyses of the AMR are frequently given in terms of the electron density-of-states diagram and the Fermi level. This is, of course, conventional in explanations of the semiconducting behavior of, say, gemanium and silicon. The Fermi level is the most probable maximum energy of the conduction electrons.

Unfortunately, however, similar analyses have little practical utility in MR materials other than for elementary pedagogical purposes. The difficulty is that the anisotropic part of the resistance depends on the exact 3-D shape of the Fermi surface, and this is not known precisely excepting for a very few magnetic materials. The Fermi surface is the 3-D envelope of the Fermi level. In some materials, the MR coefficient is negative, meaning that the low resistivity state occurs when M and I are parallel. Neither this fact nor the fact that most materials have positive MR coefficients can be easily predicted.

In the absence of a viable theory, investigators must resort to experimental work in order to find improved materials. An interesting example of such a search covered no less than 30 binary and 10 ternary alloys (McGuire and Potter, 1975). In this search, it was found that many materials, for example 80Ni/20Co, have extremely high MR coefficients (20%) at liquid helium temperatures because, of course, R is small. At room temperature, however, only the following four alloys had AMR coefficients higher than that of permalloy: 90Ni/10Co (4.9%), 80Ni/20Co (6.5%), 70Ni/30Co (6.6%), and 92Ni/8Fe (5.0%). None of these compositions is, however, suitable for MR sensors because they have either high anisotropy or high magneto-striction. The former leads to high coercive force and low permeability, the latter to high sensitivity to mechanical stress and strain.

The Unique Properties of Permalloy

It is indeed remarkable that not only were Hunt's first AMRHs made of permalloy, but also that permalloy still remains, over 30 years later, the material of choice in both conventional AMR heads and the GMR spin valve heads discussed later in this book. Moreover, permalloy is also the magnetic material of choice for most thin-film heads.

Before discussing the unique properties of permalloy that give it this dominant role in magnetic recording technology, some other important properties are listed below:

Composition (nominally)	81% Ni, 19% Fe
Saturation magnetization $4\pi M_s$	10,000 G
Electrical resistivity	20.10^{-6} Ω-cm
Magneto-resistive coefficient	2–4%
Thermal coefficient of resistivity	0.3%/°C

The first property of permalloy which is responsible for its dominance in head technology is well known. At compositions close to 81Ni/19Fe, its cubic magneto-crystalline anisotropy, K, passes through zero. Positive K means the easy axes of magnetization are the cube edges, negative K means the easy axes are the cube body diagonals, and zero K means the material is isotropic, having no preferred easy (or hard) axes.

Anisotropy is, of course, due to the spin–orbit coupling of the electron spins to the electronic structure of the material. The greater the change in spin–orbit coupling energy with changes in magnetization direction, the greater the anisotropy. It follows that in noncubic crystal structures such as hexagonal cobalt, the anisotropy is usually much larger than in cubic systems such as iron and nickel.

The fact that $K = 0$ means that in permalloy laminations and thin films produced in the absence of external "orientation" magnetic fields, the coercive force can be very low ($H_c < 1$ Oe) and the permeability very high ($\mu > 1000$). When the material is deposited or annealed in the presence of an "orientation" field, some spatial ordering of the Ni and Fe atoms occurs and the material develops what is called an *induced* uniaxial anisotropy which has its single easy axis parallel to the orientation field. The orientation field has to be sufficient to saturate the magnetization only; higher fields have no further appreciable effect. The induced uniaxial anisotropy is K_u ergs/cm^3 and is, as usual, simply the magnetic energy required, per unit volume, to rotate the magnetization from the easy axis to the hard directions orthogonal to the easy axis. In single-domain thin films, the fictitious anisotropy field, H_k, is equal to $2K_u/M$. The adjective *fictitious* is used because no actual magnetic field exists. However, the actual field required to rotate the magnetization into a hard direction is equal to H_k. In permalloy, $H_k \approx 5$–10 Oe only. The ease with which permalloy can be induced to develop an easy axis during deposition or subsequently by magnetic annealing (typically at 250 to 300°C) is a very useful property in the fabrication of both thin-film inductive heads and magneto-resistive heads.

The second unique property of permalloy is that, at a composition very close to 81Ni/19Fe, the magneto-strictive coefficient λ also passes through zero. Magneto-striction is the phenomenon in which a magnetic material changes its size depending on its state of magnetization. Figure 4.11 shows the effect. If a bar of magnetic material has length l when it is demagnetized ($M = 0$) and has length ($l + \Delta l$) when its magnetization is

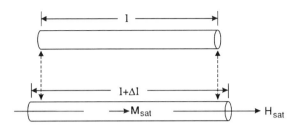

Fig. 4.11. The basic magneto-striction phenomenon showing the length of a bar changing with its magnetic state.

saturated (M_{sat}), the linear magneto-strictive coefficient, λ_l, is positive and of magnitude $\Delta l/l$. If the bar shrinks as it becomes magnetized, then λ_l is negative.

The inverse magneto-strictive effect is that the magnetic state of a magnetic material can be changed by the application of external mechanical stress. Thus a magnetized specimen with positive magneto-striction is partially demagnetized by the application of compressive stress. With negative λ_l, tensile stress is required to reduce the magnetization.

Magneto-striction is, of course, another manifestation of spin–orbit coupling. The greater the change in spin–orbit coupling energy with changes in atom-to-atom spacing, the greater the magnetostriction. It is, therefore, likely to be large when the magneto-crystalline anisotropy is large. Indeed, most cobalt alloys have large magneto-striction.

The critical importance of zero magneto-striction in head fabrication arises because it is virtually impossible to produce thin films of cubic materials which are completely stress-free. Because the top surface layer of atoms, where material is being deposited, does not have full cubic symmetry, it is inevitable that mechanically stressed films are produced. Further, differential thermal expansion effects between the several materials in the head will cause stresses. Finally, even if it were possible to deposit a stress-free head structure, it is very likely that the subsequent and inevitable mechanical processing of the head (lapping, polishing, etc.) would induce mechanical stresses.

The differential thermal expansion effects caused by I^2R heating in the copper coil of an inductive thin-film write head can give rise to magnetic disturbances when that same head cools down and contracts during subsequent reading operations. The magnetic disturbances associated with the contraction have colorful popular names such as "popcorn" or "snap-crackle" noise, and they can be minimized by careful control of the magneto-striction of the magnetic material. Because the coil heating causes tensile stresses in the magnetic yoke structure, it is generally found that making permalloy a little Ni rich, with λ slightly negative, is beneficial for maintaining the proper domain structure in the head.

The fact that, in permalloy, both $K = 0$ and $\lambda = 0$ occur at almost the same composition is responsible for permalloy's dominance in magnetic heads, both thin film and MR, in the past and in the foreseeable future.

If it is required that $K = \lambda = 0$ occurs simultaneously in permalloy, it is necessary to add a third material to the alloy. In the well known magnetic shielding material supermalloy, this is achieved by adding 4% molybdenum.

Permalloy has one notable disadvantage in that the material is mechanically soft and is thus not well suited to abrasive head-to-medium contact applications. In such applications, another $K = \lambda = 0$ material, AlFeSil, often called Sendust, is frequently used. AlFeSil is a hard, brittle alloy which resists wear well, but is considerably more demanding to produce. AlFeSil films can only be produced by sputtering, whereas permalloy films can be made by electroplating, evaporation, and sputtering.

Further Reading

Additional reading material is listed here that will prove helpful to readers who seek more detailed information.

Hunt, Robert P. (1970), "A Magneto-resistive Readout Transducer" (digest only), *IEEE Trans.* **MAG-6**, 3.

Hunt, Robert P. (1971), "A Magneto-resistive Readout Transducer," *IEEE Trans.* **MAG-7**, 1.

McGuire, T. R., and Potter, R. I. (1975), "Anisotropic Magnetoresistance in Ferromagnetic 3D alloys," *IEEE Trans.* **MAG-11**, 4.

Thomson, W. (1858), *Philosophic Magazine,* London.

CHAPTER

5

The Giant Magneto-Resistive Effect

The giant magneto-resistive (GMR) effect was unknown until 1988. It is an entirely different phenomenon from the anisotropic magneto-resistive effect treated in Chapter 4.

GMR was discovered because developments in high vacuum and deposition technology made possible the construction of molecular beam epitaxy (MBE) machines capable of laying down multiple thin layers only a few atoms thick. Such structures, which do not occur in nature, are often called superlattices because their periodicity mimics, on a large scale, crystal lattices.

Superlattices

The availability of MBE machines led to the investigation of the exchange coupling between a multiplicity of thin layers of a ferromagnet (Fe) separated by thin layers of a "nonmagnetic" material (Cr). Chromium is, of course, not actually nonmagnetic, but is an antiferromagnet when below its Néel temperature of 37°C. Typically, the layer thicknesses used were a few tens of angstroms as shown in Fig. 5.1.

The specific aspect of the interlayer exchange studied was the change from ferromagnetic (parallel) to antiferromagnetic (antiparallel) coupling between the iron layers as the chromium layer thickness is varied. The oscillatory Ruderman–Kittel–Kasuyu–Yoshida (RKKY) variation both expected and found is shown in Fig. 5.2.

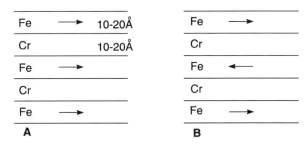

Fig. 5.1. An iron–chromium superlattice with (A) ferromagnetic (parallel) and (B) antiferromagnetic (antiparallel) exchange coupling.

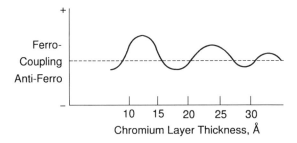

Fig. 5.2. The periodic variation of the exchange coupling in an iron–chromium superlattice versus the chromium layer thickness.

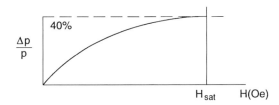

Fig. 5.3. The GMR coefficient of an antiferromagnetically coupled Fe–Cr superlattice versus field.

Measurements of the electrical resistance showed that in cases where the exchange was antiferromagnetic, the resistance could be changed as shown in Fig. 5.3, by applying large (50,000 Oe) magnetic fields. Because the resistance change was very large, $\Delta\rho/\rho_0 \approx 40\%$, this effect was called the *giant* magneto-resistive effect.

Superlattices

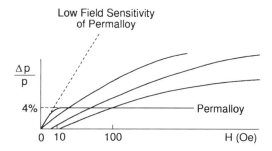

Fig. 5.4. The low-field GMR coefficient of an Fe–Cr superlattice compared to AMR in permalloy.

Studies showed that the electrical resistance was high whenever the iron layer magnetizations were antiparallel and low when they were forced, by the external field, to be parallel. The variation in resistance was found experimentally to be proportional to $-\cos \beta$, where β is the angle between the adjacent layer magnetizations. Note carefully that the GMR is independent of the measuring current direction in the plane of the films and that this is in contrast to the AMR effect, where the magnetization–current angle θ is the important factor.

Despite the excitement the discovery of GMR caused among material scientists and physicists, those involved in MRH design did not consider the effect useful for two reasons. Not only were extremely high external fields ($H_{SAT} \approx 50{,}000$ Oe) required to change the resistance, but also, as shown in Fig. 5.4, the low field sensitivity $[d(\Delta \rho / \rho_0)/dH]$ of AMR in permalloy was much higher than that displayed in the Fe–Cr superlattice GMR.

Researchers quickly realized that these shortcomings were attributable to two facts: first, the high anisotropy field ($2K/M \approx 560$ Oe) of the iron layer and, second, the use of chromium, an antiferromagnet, in the interlayers, which causes strong exchange coupling. A high external field is needed because it has to overcome both the iron layer anisotropy field and the antiferromagnetic exchange field in order to force the iron layer magnetizations to be parallel.

The solution is to use permalloy layers separated by a truly nonmagnetic metal, for example, copper, silver, or gold, as shown in Fig. 5.5. Permalloy has, as has been discussed in Chapter 4, an extremely low anisotropy field (≈ 5–10 Oe) at room temperature. Moreover, when Cu, Ag, or Au

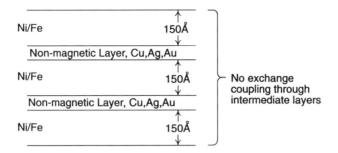

Fig. 5.5. A permalloy spin valve structure with little or no exchange coupling through the nonmagnetic intermediate layers.

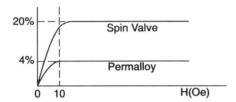

Fig. 5.6. The GMR of a permalloy spin valve compared with the AMR of permalloy.

interlayers are used there is only low RKKY exchange coupling between the permalloy layers.

Because the external field now has only to be greater than the low anisotropy field and the low exchange field, the permalloy layers can be switched from the antiparallel, high-resistance state to the parallel, low-resistance state by a relatively small field. The resistance change characteristic shown in Fig. 5.6 is typical. Note that the GMR low-field sensitivity is now larger than that found in AMR permalloy.

The precise interfacial conditions, which govern the magnitude of the exchange coupling field, are not fully known. The exchange coupling between atoms is critically dependent on their spacing, as shown in Fig. 1.4, and the RKKY coupling is an average over millions of atoms. It may well be that it is not possible to measure the interfacial structure in sufficient detail to predict accurately the exchange coupling achieved. The reasons for the imprecise knowledge of RKKY coupling are similar to those hindering the complete understanding of exchange anisotropy outlined in Chapter 6.

The Physics of the GMR Effect

Despite the novelty of GMR, simple explanations of the effect have been forthcoming. The phenomenon is believed to depend on the selective scattering of conduction electrons at the interfaces between the ferromagnetic and nonmagnetic layers and also the differing penetration depths of these electrons in the ferromagnetic layers.

The conduction electrons in iron and permalloy are the $3d$ electrons discussed in Chapter 1. Every electron has the quantum-mechanical property called spin. Equation (1.7) shows the majority of the electron spin directions in the ferromagnet are oriented parallel to the magnetization vector M.

When an electric field is applied, the spin-oriented conduction electrons accelerate until they encounter a scattering center. The existence of scattering centers is, of course, the origin of the electrical resistivity. The average distance the conduction electron accelerates is called the *coherence or penetration length*, and it is of the order of 50–100 Å in the metals used in GMR superlattices.

Consider the case in which the scattered electron traverses the interlayer of a nonmagnetic metal. Provided the interlayer thickness is less than the coherence length, the conduction electron arrives at the interface of the adjacent ferromagnetic layer still carrying its original spin orientation. This process is shown in Fig. 5.7.

When the adjacent magnetic layers are magnetized in a parallel manner, as in Fig. 5.7A, the arriving conduction electron has a high probability of entering the adjacent layer with negligible scattering, because its spin orientation matches that of the adjacent layer's majority spins. Moreover, the penetration depth of the electrons is relatively large, say λ_1.

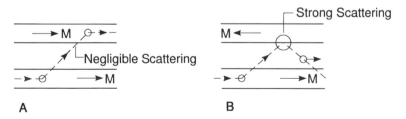

Fig. 5.7. The scattering of a spin-polarized, conduction electron at the interface of (A) a ferromagnetically and (B) an antiferromagnetically coupled layer.

On the other hand, as shown in Fig. 5.7B, when the adjacent layer is magnetized in an antiparallel manner, the majority of the spin-oriented electrons suffer strong scattering at the interface because they do not match the majority spin orientation and the penetration depth is small, say λ_2. When the magnetic layers are in the ferro state (magnetized parallel) the resistance is low, and vice versa in the antiferro state (magnetized antiparallel).

The precise details of the interface which controls the magnitude of the strong scattering are not known in detail. It is known that the surface roughness of the interface is critical. As measured by low-angle X-ray diffraction, roughnesses of a few angstroms give the greatest scattering and, thus, highest GMR. Bearing in mind that atom-to-atom spacings are in the range of 2 to 3 Å in these materials and the fact that X-ray diffraction measurements are averages of the roughness over the X-ray beam projected area (typically, many square micrometers and millions of atoms), we may well wonder just precisely what a measured roughness of 2 to 3 Å really means. Once more, as in RKKY exchange coupling at interfaces, the atomic details necessary for the development of high GMR may be beyond experimental determination. Similarly, there is no precise understanding of the penetration depths λ_1 and λ_2.

Despite the uncertainty in the exact interfacial structure required, it is, however, possible to list several necessary conditions for the development of GMR.

First, the two materials used in the superlattice must be immiscible. In other words, they must not mix or dissolve in each other, just as oil and water are not soluble in each other. If the materials are miscible, the interfaces, and eventually the interlayers, can be expected to diffuse into each other over time and thus become no longer sharply defined. The well-known two criteria for the immiscibility of two materials are that their atom–atom spacings must differ by more than 15% or their crystal habits must differ.

The next prerequisite for GMR is that the ferromagnetic layers must have some mechanism, whether exchange coupling or mere magnetostatics, that establishes the antiferro (antiparallel) magnetization state in zero external field.

Third, the interlayer of nonmagnetic material must be thinner than its conduction electron coherence length. This is so that the electrons can "remember" their spin orientations as they cross between the

The Physics of the GMR Effect

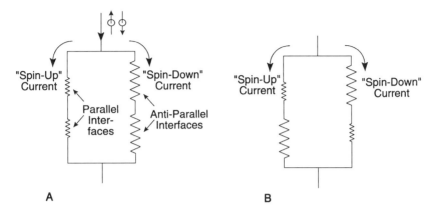

Fig. 5.8. The two-current model electrical circuit showing (A) the low-resistance and (B) the high-resistance conditions.

ferromagnetic layers. With coherence lengths of 50–100 Å, this means that nonmagnetic interlayer thicknesses no more than a few tens of angstroms are required.

In each ferromagnetic layer, two spin orientations of the conduction electrons coexist. The majority are parallel to the magnetization and the minority antiparallel. The magnetization is just the difference between the numbers of majority and minority electrons, per unit volume. This fact has led to the idea of a "two-current" GMR model.

The two-current model is shown in Fig. 5.8. It is simply a two-leg, four-resistor parallel network. In the low-resistance state shown in Fig. 5.8A, the opposite legs of the parallel circuit have two low and two high resistances, respectively. The two currents, electron spin-up and spin-down, enter at the top. The spin-up current flows down the left leg, which has two low resistances corresponding to two parallel magnetization interfaces. The spin-down current flows through the two high resistances in the right leg that models the antiparallel interfaces.

In Fig. 5.8B, the high-resistance state of the two-current model is shown. Now, in whichever leg the two spin-oriented currents flow, they encounter equal resistance. There is no easy, low-resistance leg of the circuit for either the majority or minority electrons.

By putting the parallel and antiparallel layer resistances to be equal to R_{LO} and R_{HI}, respectively, it is easy to show that the difference in resistance, ΔR, between the high and low resistance conditions of the two

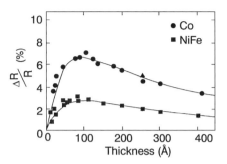

Fig. 5.9. The variation of the GMR coefficient at room temperature with the ferromagnetic layer thickness.

current circuit is proportional to $(R_{HI} - R_{LO})^2$. This shows that a high GMR effect depends upon a large difference between R_{HI} and R_{LO}. Alternatively, by putting $R_{HI} \propto 1/\lambda_2$ and $R_{LO} \propto 1/\lambda_1$, ΔR is proportional to $(\lambda_1 - \lambda_2)^2$, showing that a large difference in the penetration depths of the majority and minority carriers leads to high GMR.

The ratio λ_1/λ_2 has been measured for several ferromagnetic metals of interest: permalloy, 8; iron, 2; and cobalt, 6. It will be noted that, amazingly enough, permalloy gives the best GMR. Typically, GMR coefficients, $\Delta R/R$, of the materials in the spin valve heads discussed in this book are in the range 5–10% at room temperature.

Figure 5.9 shows the variation of GMR coefficient, $\Delta R/R$, versus ferromagnetic layer thickness at room temperature for cobalt and permalloy. It is seen that when the thickness, d, is less than the minority penetration or coherence length, λ_2, there is, as expected, very little GMR. When d is between the minority, λ_2, and majority, λ_1, penetration lengths, the GMR increases rapidly with increasing d because the ratio of the penetration depths is increasing. Finally, when d is greater than the majority length λ_1, the GMR falls off with increasing d because the GMR effect is increasingly being electrically shunted with an inactive layer which has no GMR effect.

The variation of the GMR coefficient with temperature is shown in Fig. 5.10 for cobalt and permalloy. In both cases, $\Delta R/R$ increases with decreasing temperature. As is the case in AMR, these changes occur because R increases at lower temperatures with ΔR, with the magnetoresistive change of resistance being almost independent of temperature.

As is the case in explanations of AMR which invoke the statistics of the electron energies, band structure, and the density of states, similar

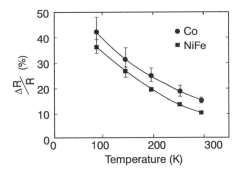

Fig. 5.10. The variation of the GMR coefficient of cobalt and permalloy with temperature.

explanations of GMR are frequently put forth. Such arguments are common in pedagogical expositions. However, they lack real utility because they do not include the structural detail of surface roughness and differing coherence lengths that is known to be necessary to explain GMR phenomena.

Granular GMR Materials

Shortly after the discovery of GMR in superlattices, analogous discoveries were made in granular, or heterogeneous, GMR materials.

The first such material was made by sputtering alternating, relatively thick layers of cobalt and copper. Such a layered structure of two immiscible metals is in thermodynamic equilibrium because the high-energy interfacial area is minimized. Recall that after vigorous shaking of oil and water, the small oil droplets rapidly coalesce into layers in order to reduce the oil/water interfacial area and energy.

The sputtered multilayer of cobalt and copper was then annealed, during which process the layers of cobalt diffuse to become irregular particles or grains in a matrix of copper as shown in Fig. 5.11. An optimum annealing time and temperature exists (about 300°C for 1 hour) that produces the desired microstructure where the cobalt grains are separated by distances less than the coherence length of electrons in copper.

Unfortunately, the low field sensitivity $d(\Delta\rho/\rho_0)dH$ of these first granular GMR materials was much too low for MRH device applications. Again the problem was in the improper selection of the ferromagnetic material, cobalt, which has an extremely large anisotropy field ($2K/M \approx$ 6000 Oe). Moreover, the cobalt grain sizes achieved after the annealing

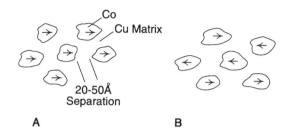

Fig. 5.11. The magnetization state of a granular cobalt–copper two-phase GMR material in (A) the low-resistance and (B) the high-resistance states.

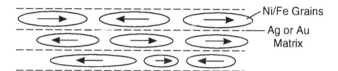

Fig. 5.12. A magnetization state of a premalloy–nonmagnetic metal two-phase GMR structure in its demagnetized (zero external field) state.

were much too small, being well into the superparamagnetic range (a few tens of angstroms). For these two reasons, large external fields (≈50,000 Oe) were needed to force the cobalt grains into the parallel magnetization state.

The solution to the low-sensitivity problem was obvious: replace the cobalt with permalloy and make the grain sizes larger. Accordingly, in the most recent granular GMR materials, the starting point is a sputtered multilayer of permalloy and silver. Unlike copper, silver is not miscible in permalloy. After suitable annealing (again at 300°C for 1 hour) the silver diffuses down the grain boundaries (but not into the grains themselves), resulting in the structure shown in Fig. 5.12.

Now, because the low anisotropy permalloy grains are not superparamagnetic, the required saturation external magnetic field is of the order of 10 Oe only, and a large low-field GMR sensitivity can be realized. Granular GMRHs have, however, proved to have disappointing performance. This is principally due to the fact that the high resistance state, shown in Fig. 5.12, is not well defined and, even worse, not exactly reproducible. This demagnetized state occurs solely as a consequence of the reduction of magnetic self-energy, and a multiplicity of such states of low energy exists. Accordingly, the granular GMRHs appear to be noisy in operation and have, therefore, low signal-to-noise ratios. Although granular GMR continues as an illuminating and entertaining field for academic research, it is believed that no granular GMR head work has occurred since 1995.

It is important to realize that both the superlattice and granular GMR materials can now be made without the requirement for ultrahigh-vacuum, expensive MBE machines. The fact that these materials can be made with ordinary laboratory sputtering equipment is surely one of the most important reasons for the widespread application of GMR (spin valve) heads today.

Further Reading

Additional reading material is listed here that will prove helpful to readers who seek more detailed information.

Baibich, M. N., Broto, J. M., Fert, A., Nguyen Van Dau, F., and Petroff, F. (1988), "Giant Magnetoresistance of (001)Fe/(001)Cr Magnetic Superlattices," *Phys. Rev. Lett.* **61**, 21.

Dieny, B., Speriosu, V. S., Parkin, S. S. P., Gurney, B. A., Wilhoit, D. R., and Mauri, D. (1991), "Giant Magnetoresistance in Soft Ferromagnetic Multilayers," *Phys. Rev. B* **43**, 1.

Hylton, T. L., Coffey, K. R., Parker, M. A., and Howard, J. K. (1993), "Giant Magnetoresistance at Low Fields in Discontinuous NiFe-Ag Multilayer Thin Films," *Science,* p. 261.

White, R. L. (1992), "Giant Magnetoresistance — A Primer," *IEEE Trans.* **MAG-28**, 5.

CHAPTER

6

Vertical Biasing Techniques

In the design of both AMRHs and GMR spin valves it is usually necessary to provide both vertical and horizontal magnetic bias fields. In this chapter most of the techniques proposed for applying a vertical bias field on a magneto-resistive element are discussed in historical order. As will be evident by noting the very number of techniques that are treated here, this has been an extraordinarily fertile field for invention.

Vertical bias is sometimes called perpendicular bias and, rather confusingly, transverse bias. Horizontal bias, also known as longitudinal bias, is discussed in Chapter 7.

The description of each biasing technique is followed by a short review of its main advantages and disadvantages.

Permanent Magnet Bias

The earliest, and one of the simplest, ideas for providing vertical bias is shown in Fig. 6.1. Here a thin film of permanent magnet (PM) material is deposited adjacent to the MR element. When the permanent magnet is magnetized up, it produces a down bias field in the MRE. The coupling is purely magneto-static. To prevent exchange coupling and electrical shunting of the measuring current and output voltage by the PM film, a nonmagnetic, electrically insulating film is interposed as shown. The insulator film is typically 100 to 200 Å of SiO_2 or Al_2O_3.

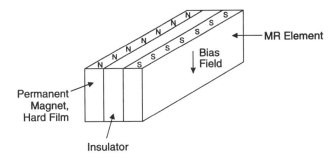

Fig. 6.1. The permanent magnet bias structure.

The main advantage of this scheme is that the many years of experience in making thin-film recording media can be applied to the PM film. In the past, PM films have been made of Co–P or Co–Ni–P, whereas Co–Cr–Pt–Ta is common today. The design problem is also relatively simple. The PM film must have a remanence times thickness product that is approximately equal to $M_s T \sin\theta$ of the MRE.

Disadvantages include the fact that a high magnetic field, of magnitude about equal to the coercivity of the PM material, is present at the top edge immediately adjacent to the tape or disk. It follows that the PM film coercivity must be limited to only 50% of that of the recording medium. Another disadvantage is that the errors in fabrication, which result in an incorrect vertical bias field, cannot be adjusted easily after the MRH has been made. Postfabrication adjustment of the vertical biasing is, on the other hand, easily accomplished in the current biasing schemes discussed next.

Current Bias

In this early scheme the current flowing into the MRH is divided into two parts. One flows through the MRE as usual. The second part flows through a layer of electrically conductive, nonmagnetic material which is electrically insulated from the MRE. The arrangement is shown in Fig. 6.2.

The current flowing in the MRE cannot provide a vertical bias field in the MRE itself. However, the current in the adjacent layer does cause a vertical bias field in the MRE. By the right-hand rule discussed in Chapter 1, current flowing in the direction shown in Fig. 6.2 produces a down vertical bias field in the MRE.

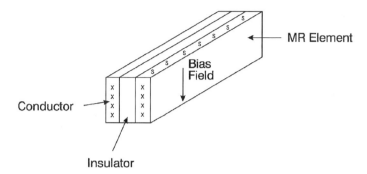

Fig. 6.2. The current bias arrangement.

The magnitude of the current is determined by the voltage applied across the MRH and the resistance of the current conductor film. If the shunt current is I_s and the conductor is of equal depth to the MRE, then the vertical bias field is, over the middepths of the MRE, approximately equal to $0.2\pi I_s/D$ Oe. An estimate can be made of the MRE bias angle using Eq. (4.4).

An advantage of current bias is that it requires only minimal materials technology to implement. Moreover, by adjusting the voltage applied across the MRH, the current and thus the vertical bias field can be adjusted after head fabrication.

The main disadvantage of current bias is that the effective resistance change is reduced by the electrical shunting when the MRH is operated on a two-terminal device. The effective resistance change is $\delta R R_2/(R_1 + R_2)$ only, where R_1 and R_2 are the MRE and conductor resistances, respectively. In some designs the shunt conductor has been made of titanium, in order to increase R_2. Of course, if the MRH is operated as a three-terminal device, the full MRE resistance change δR can be detected at the expense of increased circuit complexity.

A second disadvantage is that vertical bias field magnitude falls off to the approximately $0.1\pi I_s/D$ at the top and bottom ends of the MRE. The critical region of the MRE which is closest to the recording medium is likely to be underbiased. The depth of this underbiased region can be reduced by making the insulator film thinner or by omitting it altogether. When no insulator film is used, however, problems can arise due to diffusion of the conductor layer metal into the permalloy MRE. For example, copper is soluble in permalloy.

Soft Adjacent Layer Bias

The soft adjacent layer (SAL) vertical bias arrangement is shown in Fig. 6.3. Here a film, made of a low-coercivity, high-permeability magnetic material, is placed adjacent to the MRE. To prevent electrical shunting, a thin layer of insulator is interposed.

Typically, the soft film is permalloy, but some designs use other materials, such as AlFeSil or NiFeCr. Insulator layers of SiO_2 or Al_2O_3 and AlN are used today, with a high resistivity (tetragonal) phase of Ta being common in many of the older hard-disk file MRH designs.

The principle of operation of the SAL vertical bias technique is as follows. The current flowing in the MRE produces, by the right-hand rule, a vertical magnetic field in the SAL. This field magnetizes the SAL in the vertical direction. In Fig. 6.3, the MRE current direction shown causes the SAL to be magnetized up. The changing SAL magnetization produces $\rho = -\nabla \cdot M$ magnetic poles, as was discussed in Chapter 1, and these poles generate the desired down vertical bias field in the MRE. Because there is mutual magneto-static interaction between the SAL and the MRE, a simple and accurate way of calculating the vertical bias field in SAL MRHs is not known to this writer. Accordingly the design problem must be solved by using micromagnetic numerical techniques.

The design is usually arranged so that the middepth region of the SAL becomes magnetically saturated. This is done because saturation of the SAL reduces its permeability, which, in turn, reduces the undesirable record medium fringing flux shunting that is inherent in all SAL MRHs.

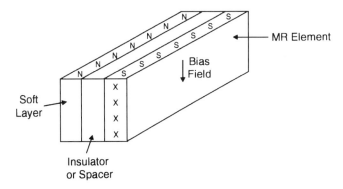

Fig. 6.3. The SAL bias construction.

This signal flux shunting occurs because there are two parallel paths for the record medium fringing flux, the SAL and the MRE.

The advantages of SAL MRHs include the fact that adjustments to the vertical bias field can be made by adjusting the current. In principle, the materials technology requirements are minimal.

The SAL designs have two main disadvantages. The first, the signal flux shunting problem, has already been discussed. The second is that correct operation of the head depends on the integrity of the electrical insulating film. If pinholes occur in fabrication or if the SAL and MRE become wiped or smeared together by tape or disk abrasion, the performance is compromised by electrical shunting.

The SAL technique is used by IBM in current tape drives and in older (pre-1997) disk file AMRHs.

Double Magneto-Resistive Element

The double-element MRE vertical biasing scheme is a natural evolution of the current bias and SAL bias techniques already discussed.

As shown in Fig. 6.4, the design has two MREs separated by a thin insulator layer. Both MREs carry current, which usually, but not necessarily, flows in the same direction. Current in MRE #1 produces an up vertical bias field in MRE #2. Current in MRE #2 produces a down field in MRE #1.

Because the two MREs interact mutually, a simple way to estimate accurately the magnetization bias angles does not exist.

Advantages include the minimal materials science required, the simple fabrication, and the adjustability of the vertical bias by changing the

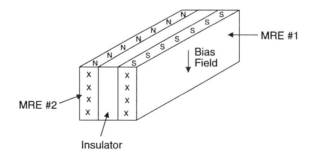

Fig. 6.4. The double-element magneto-resistive bias technique.

current magnitudes. Moreover, the media flux shunting problem of the SAL design is avoided because a changing magneto-resistive output signal is available from both MREs.

The two MRE outputs can be sensed either in addition, as in the Kodak Dual-Magneto-Resistor (DMR) head, or in subtraction, as in the Hewlett-Packard Dual-Stripe head. The relative advantages and disadvantages of these arrangements are discussed in Chapter 15.

Self-Bias

This particularly ingenious vertical biasing technique is applicable to MRHs in which the MRE is placed in the gap between two highly permeable poles or shields. The function of shields in MRHs is covered in detail in Chapter 9.

Figure 6.5 shows the MRE placed off-center between shields #1 and #2. The off-center positioning is the essence of the invention.

When current flows in the MRE, it of course cannot produce a vertical bias field on itself directly. To meet the magnetic boundary conditions imposed by the high permeability shields, however, a vertical bias field does appear when the MRE is not on the centerline of the gap.

Referring to Fig. 6.5, the effect of the shields' presence is to alter the field produced by the MRE current in a way that looks, in the gap, as though magnetic "images" exist in the shields. In first approximation, these images look like the MRE itself, just as one's own image in a mirror looks like one's self. The images appear to exist at a distance in the shields equal to the MRE-to-shield spacing, just as your own reflection in a mirror appears to be as far behind the glass as you are in front.

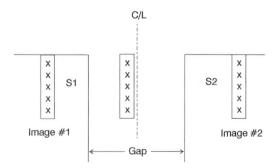

Fig. 6.5. The self-bias method.

Because distance from the MRE to the left-hand image (#1) is smaller than the right-hand image (#2) distance, the vertical bias fields (H_1 and H_2) that each image produces at the MRE are not equal. The net vertical bias field in the case shown in Fig. 6.5 is of magnitude $H_1 - H_2$ and is down.

The main advantage, of course, of self-bias is that it occurs automatically, merely by putting the MRE off-center. No extra structure is required.

The main disadvantage appears to be that the magnitude of the vertical bias field is often not sufficient. This insufficiency arises because the images are too far away from the MRE and the net bias field is the difference, $H_1 - H_2$. Very often the self-bias vertical bias is augmented by using one of the other techniques discussed in this chapter.

Exchange Bias

Exchange bias, shown in Fig. 6.6, depends on atomic contact between the MRE and another magnetic material. Atomic contact means that every effort is made to ensure that no foreign material or atoms exist in the interface. When the proper interface is achieved, the quantum-mechanical exchange phenomenon, discussed in Chapter 1, couples the adjacent electron spins in the MRE and the other magnetic layer.

The interfacial exchange coupling energy depends on the submicroscopic details of the interface. The strength of the coupling can be deduced by analysis of the *M–H* loop characteristics of the coupled films. It is usually expressed as an effective exchange field, H_{EX}. When the interfacial

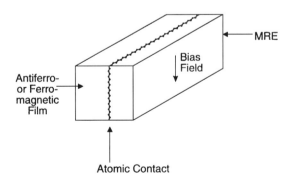

Fig. 6.6. The exchange bias principle.

exchange coupling energy is J_{EX} ergs·cm^{-2} and the MRE magnetization and thickness are M_s and T respectively, the effective exchange field is $J_{EX}/M_s T$.

The interfacial exchange field can have a profound effect upon the M–H loop characteristics of the coupled films. Suppose that the antiferromagnetic layer has magnetocrystalline anisotropy, K ergs·cm^{-3}, and thickness, t.

When $J_{EX} < Kt$, the M–H loop is shifted unidirectionally because the interfacial coupling is too weak to reverse the spins throughout the entire thickness of the antiferromagnet, and this result is called *exchange anisotropy*. When $J_{EX} > Kt$, the effect is principally that of increasing the coercivity.

Antiferromagnetic films of MnFe, CoO, and TbCoFe and ferromagnetic films of CoP, CoNiP, and CoCrPt are used in MRH biasing. The ferromagnetic films may or may not be exchange coupled depending on whether atomic contact has been established. The exchange-coupling field of a ferromagnetic film is of opposite polarity to that film's magnetostatic coupling field.

The magnitude of the exchange field varies widely from a few oersteds to several hundred oersteds. The exact control of the exchange field is a very difficult materials science problem because it depends on atomic details of the interface that are, by their very nature, almost unmeasurable. The magnitude and sign of the exchange coupling between a pair of atoms is a rapidly varying function of the atom-to-atom spacing, as is shown in Fig. 1.4. The measured exchange field is the coupling averaged over millions of interfacial atoms.

When an antiferromagnetic film is used, the bias field generated cannot be reset or changed accidentally by magnetic fields alone. To change the antiferromagnetic exchange anisotropy direction, it is necessary to cool the antiferromagnet from above its Néel temperature in the presence of a magnetic field. The Néel temperature is the antiferromagnetic analog of the ferromagnetic Curie temperature.

When a ferromagnetic exchange bias is used, the exchange bias direction can be reset or changed merely by using a magnetic field sufficiently greater than the coercivity of the ferromagnet.

A disadvantage of exchange bias, other than the difficult materials science and fabrication technology already discussed, is that many of the commonly used antiferromagnets are prone to oxidation and corrosion. Exchange bias is used in all GMR spin valve heads and a wide variety of antiferromagnetic materials have been studied, as may be seen in Table 11.1 in Chapter 11.

The Barber-Pole Scheme

In the barber-pole biasing scheme, an entirely different approach is taken to the fundamental vertical biasing problem of establishing the correct angle between M and I. As shown in Fig. 6.7, instead of using magnetic bias to rotate the MRE magnetization M to the appropriate angle θ, the current is made to flow obliquely in the MRE.

The permalloy MRE is overcoated with a thick conductor film that is photolithographically processed to etch out the "barber's pole" pattern shown. By making the conductor layer thick, essentially no current flows in the MRE in the regions below the conductors. Between the conductor stripes, however, the current must flow in the permalloy and it does so in the shortest possible path, that is, orthogonal to the conductor stripes. The result is that the proper angle between M and I is imposed by the geometry of the stripes.

In some designs, the stripes are arranged in a chevron pattern so that on one side of the MRE the M, I angle is positive and on the other it is negative. This elaboration is undertaken to reduce the asymmetry in the positive and negative amplitude response inherent in a single-element MRE. Pulse amplitude symmetry can also be obtained in the split-element and servo-bias schemes discussed next and in some of the double-element MRE designs discussed in Chapter 15.

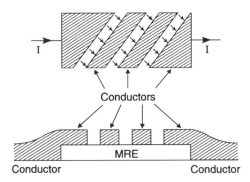

Fig. 6.7. The barber-pole structure.

When the barber-pole idea is simplified to its irreducible minimum, the M, I current angle θ can be established with merely a pair of "slant conductors." This technique has increasing appeal as the track width and MRE width W is decreased.

The main advantage of barber-pole bias is that the M, I angle is set directly by photolithography. There are no magneto-static (demagnetizing field or adjacent layer interaction field) complications.

Among the disadvantages is the unfortunate fact that parts of the MRE are, by design, not used to sense the signal flux from the recording medium. Estimates have been made that no more than about 60% of the MRE width W can be used as an active MR sensor.

The barber-pole vertical bias scheme is used in the Philips Digital Compact Cassette (DCC) audio recorder's yoke-type MRHs. Yoke-type MRHs are discussed in Chapter 14.

The Split-Element Scheme

The split-element idea, shown in Fig. 6.8, is not really a means of producing vertical bias fields. Rather, it is a clever method of minimizing the signal asymmetry or distortion that results when the vertical bias field, applied by one of the other methods, is not of the correct value.

By providing a center tap, the current in the MRE is made to flow in opposite directions on either side. This is shown in Fig. 6.8.

Suppose that a vertical bias field has been produced by one the methods using the MRE electrical current, for example, self-bias. Because the

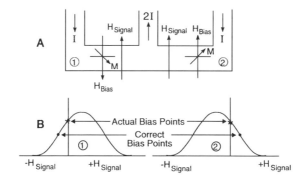

Fig. 6.8. (A) The split-element arrangement and (B) the push–pull idea.

current flows in opposite directions in either side, the polarity of the vertical bias field is opposite in either side. In Fig. 6.8(A) the down vertical bias field is shown on the left-hand side and vice versa on the right.

Suppose that the magnitude of the vertical bias fields is not correct. In Fig. 6.8B the variable resistance versus signal field H_{signal} characteristics of the two sides are shown with both being underbiased. To achieve small-signal linearity with minimal pulse amplitude asymmetry, the MREs require a greater magnitude of vertical bias field.

When the split-element senses the fringing field of the recording medium, H_{signal}, the changes of resistance are of opposite sense on either side. For the up signal field shown in Fig. 6.8A, the resistance of the left side increases, whereas that of the right side decreases.

These resistance changes are sensed differentially. The differential sensing cancels out the pulse asymmetry due to the incorrect vertical biasing. In fact, the technique of differentially sensing the opposite polarity outputs of two devices is called "push–pull" operation in electronic design, and it is extremely common in, for example, high-power audio amplifiers.

The advantage of the split-element idea is that the differential sensing improves the pulse amplitude symmetry due to improper vertical biasing, regardless of whether that bias is too small or too great. The scheme is, therefore, appealing in analog applications, such as audio recording, where the allowable (second) harmonic distortion is much lower than in the case in digital (binary) recording.

The main disadvantage is, apart from the circuit complexity, the fact that the portion of the MRE underneath the center conductor cannot sense signal because it has neither a bias field nor a measuring current. The split-element arrangement is used in early IBM 0.5-in. tape drive AMRHs.

The Servo-Bias Scheme

Another innovation in vertical bias technology is shown in Fig. 6.9. It is an evolution of the current bias scheme already discussed. In the servo-bias scheme, the current in the adjacent conductor must be electrically isolated from that in the MRE.

The current flowing in the adjacent conductor shown in Fig. 6.9 produces a down vertical bias field in the MRE. However, this current is not generated by a constant current (high-impedance) source, but rather from a variable current driver.

The Servo-Bias Scheme

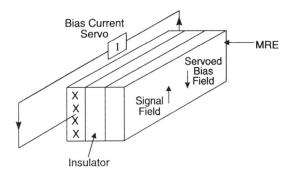

Fig. 6.9. The servo-bias system.

The variable driver is, in fact, a negative feedback electronic circuit whose output is the variable bias current. The error signal fed back to the bias current servo is derived from the magneto-resistive variable change, δR, of the MRE. The actual error signal could be, for instance, the total even harmonic distortion of the MRE voltage.

Thus the function of the bias current servo is to adjust automatically the bias current to the proper amplitude which gives the minimum even harmonic distortion. This value of the bias current produces precisely the correct vertical bias field in the MRE. It should be clearly understood that the precision which can be attained with automatic servoing of the vertical bias is limited only by the signal-to-noise ratio available in the negative feedback error signal.

Two modes of servo-bias operation can be visualized. In the first, the servo-bias current driver has only a low-frequency response or bandwidth. In this case the signal output is just the usual varying resistance of the MRE. In this mode, the servo-bias merely ensures that the MRE remains at the optimum bias point which minimizes, say, pulse amplitude asymmetry or even harmonic distortion.

In the second mode, the servo-bias driver has a much greater bandwidth than that of the recorded signals. Now the voltage drop across the MRE can be held fixed and the varying servo-bias current becomes the signal output of the MRH. In this case, subject only to the signal-to-noise ratio of the negative feedback error loop, the variable servo-bias current and output signal become almost free of harmonic distortion. Moreover, because the MRE is now simply a "null" detector, the actual slope of its variable resistance versus field characteristic is no longer of great importance!

The advantage of the servo-bias scheme is particularly obvious in multitrack MRHs where the problem of achieving uniformity in the resistance change versus current characteristic of each track is extremely difficult. The production yield of multitrack MRHs can be increased greatly by the larger track-to-track variations that are tolerable when servo-bias is used.

The disadvantages of servo-bias are equally obvious. A relatively complicated bias current servo circuit must be provided. The bias current servos currently used are implemented in a large-scale integrated (LSI) silicon chip.

Servo-bias is used in the Philips DCC audio recorders. In the fixed DCC heads, eight analog tracks are read by using a pair of the digital MR read heads, servo-biased together. In the rotary DCC heads, each analog track is servo-biased independently.

As an aside, this writer cannot help noting that despite the fact that Bob Hunt invented AMRHs at Ampex, no practical application for AMRHs could be found in the analog audio, instrumentation, and video recorders made by Ampex in the period 1960–1990. The principal reasons for this failure to find applications were twofold. First, due to the photolithographic tolerances, each individual head requires a slightly different vertical bias current for optimum operation. Second, even at the optimum point, the dynamic range (the range between maximum and minimum analog signal levels) was always too small, being limited by the harmonic distortion inherent in AMRHs. Proposals made at Ampex, by this writer, to use a vertical servo-bias system were, in that time frame, deemed to be "obviously too complicated" or "too expensive" to implement. In the 1990s, however, no fewer than eight pairs of servo-bias MRHs were operated simultaneously in a portable consumer audio entertainment machine!

Further Reading

Additional reading material is listed here that will prove helpful to readers who seek more detailed information.

Bajorek, C. H., Krongelb, S., Romankiw, L. T., and Thompson, D. A. (1974), "An Integrated Magnetoresistive Read, Inductive Write High Density Recording Head," in *20th Annual AIP Conf Proc.,* American Institute of Physics, p. 24.

Further Reading

Kuijk, K. E., van Gestel, W. J., and Gorter, F. W. (1975), "The Barber Pole, A Linear Magnetoresistive Head," *IEEE Trans.* **MAG-11**, 5.

Markham, David, and Jeffers, Fred (1991), "Magnetoresistive Head Technology," *Proc. Electrochemical Soc.* **90**, 8.

Shelledy, F. B., and Nix, J. L. (1992), "Magnetoresistive Heads for Magnetic Tape and Disk Recording," *IEEE Trans.* **MAG-28**, 5.

CHAPTER

7

Horizontal Biasing Techniques

Horizontal biasing is necessary in all MRHs so that the single-domain magnetization state of the MRE or sensor film will be stable against all reasonable perturbations. These perturbations include external magnetic fields and thermal and mechanical stresses.

As indicated in Fig. 7.1, it is almost inevitable that the end zones of the MRE will not be in the single-domain magnetic state, but that the magnetization will break up into the multidomain state. If the multidomain state did not occur in the end zones, the demagnetizing field in those zones would be extremely high ($H_d \approx 4\pi M_s \cos\theta$). In reality, of course, the demagnetizing field in the end zones cannot exceed the coercivity of the MRE because domain walls are then nucleated and a multidomain state is created.

The primary function of the horizontal bias is to keep the MRE in the single-domain state. Whenever domain walls move beyond a small reversible range, they jump irreversibly to new equilibrium positions. These irreversible "Barkhausen" domain wall jumps are the origin of hysteresis in soft, that is, low-coercive-force, magnetic materials.

When the horizontal bias is not adequate to keep the MRE single domain, domain wall jumps change the demagnetizing field in the MRE. This changes the magnetization versus current angle and changes the variable resistance versus field characteristic curve. Because domain wall jumps show hysteresis, the resistance characteristic curve also displays hysteresis. Hysteresis in the resistance characteristic means that the increasing field curve is no longer the same as that for decreasing fields.

Conductor Overcoat

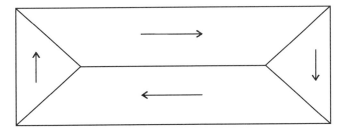

Fig. 7.1. The lowest energy state of an MRE showing the multidomain end zones.

Many of the techniques used for producing vertical bias have been proposed for horizontal bias. Again, the field is replete with novelty and ingenuity.

Clearly, the permanent magnetic (PM) film arrangement shown in Fig. 6.1 could be used for applying both vertical and horizontal bias fields by merely setting the PM film's magnetization at an oblique angle. Similarly, the exchange bias scheme shown in Fig. 6.6 could be used for both vertical and horizontal biasing. On the other hand, the vertical bias schemes that use electric currents cannot provide horizontal bias fields.

The horizontal biasing arrangements discussed in detail in this chapter are those which are not merely simple adaptations of vertical bias methods. Closely allied to horizontal bias techniques are other methods used to minimize the actual effect of unstable domain walls in the end zones. An outline of these methods is also included in this chapter.

The greater the horizontal bias, the lower becomes the MRH's sensitivity, and it follows that the bias is made sufficient only for perturbing fields which might reasonably be expected. Typically a horizontal bias field of about 100–150 Oe is used.

Conductor Overcoat

Perhaps one of the simplest ideas is shown in Fig. 7.2. Here the MRE is simply made much wider than the required track width, W. The troublesome multidomain end zones of the MRE are arranged so that they do not carry any magneto-resistance measuring current. In Fig. 7.2 this is achieved by overcoating the end zones with a sufficiently thick conductor layer. By allowing little current to flow in the end zones, the end zones contribute little to the MRH output signal.

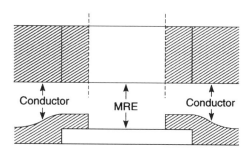

Fig. 7.2. The conductor overcoat technique.

Moreover, in the single-channel MRHs used in disk files, the end zones can be almost arbitrarily distant from the center current-bearing sensing region of the MRE and thus the effect of the end-zone domain wall instability can be made very small. A further elaboration on this idea is to progressively thin the MRE thickness T toward the end zones. This thinning raises the coercivity of the film and helps to pin the domain walls in place.

Another advantage of conductor overcoat is that the precise MRE sensing track width W is set directly by the photolithographic process. Furthermore, by careful design of the way the current leads join the conductor overcoat, useful horizontal bias fields can be generated in the end zones.

Exchange Tabs

Another approach to pinning the end-zone domain walls is shown in Fig. 7.3. Here, the end zones are exchange coupled to either a ferromagnetic or antiferromagnetic layer. Exchange coupling was discussed in detail in Chapter 6.

When the exchange coupling field is sufficiently large, the exchange tabs fulfill two roles. First, any end-zone domain walls are pinned, and thus the MR characteristic displays no hysteresis. Second, even though measuring current is flowing through the MRE under the exchange tabs, no magneto-resistive signal is generated because the magnetization versus current angles cannot change in the pinned region.

Originally, exchange tabs were made with MnFe, as was the case in the first two generations of IBM disk file AMRHs.

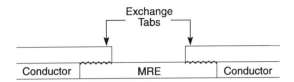

Fig. 7.3. The exchange tab arrangement.

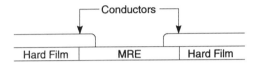

Fig. 7.4. The hard-film geometry.

Hard Films

Figure 7.4 illustrates yet another approach to maintaining the single-domain magnetization state. Films of high coercivity material are abutted against the ends of the MRE. We can understand how this works by thinking about it in two ways. First, by arranging for the flux flow ($B_s \cos \theta\ TD$) of the MRE to be equal to the flux flow of the hard film, we see that, because no magnetic poles exist at their junction, the demagnetizing field in the end zone vanishes. Alternatively, we can imagine that the magnetic poles at the ends of the hard film produce a sufficient horizontal bias field in the MRE.

In Fig. 7.4, the precise sensing track width of the MRH is defined by the current conductors.

An obvious practical difficulty in the implementation of hard-film horizontal bias arises. The efficacy of the scheme depends strongly on the microscopic details of the MRE–hard-film junction. This problem can be mitigated, but not eliminated, by sloping the interface so that the actual junction is spread over a distance as large as perhaps 0.1 μm.

Hard-film horizontal biasing replaced exchange bias in later generation IBM hard-disk file AMRHs. Moreover, in these designs "hard bias" is applied to both the MRE and the SAL as is shown in Fig. 9.13. All GMR spin valves use "hard bias" on the sensor film.

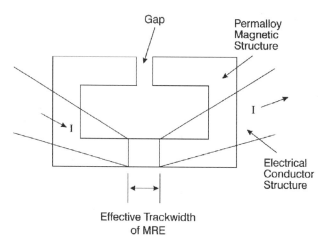

Fig. 7.5. The closed-flux structure.

Closed Flux

In multichannel MRH assemblies, use is made of yet another idea for reducing the effect of the end-zone domain wall instability. Figure 7.5 shows an MRE in which the end zones have been folded around above a conductor defined sensing region. The two end zones are separated by a small gap in order to prevent electrical shorting of the MRE. The entire structure shown is made of permalloy.

Recall that a closed structure, such as a toroid, can be magnetized circumferentially without generating $\rho = -\nabla \cdot M$ magnetic poles and, therefore, both the internal (demagnetizing) or external (fringing) fields are zero. A divergence-free magnetization state is called a *closed-flux* pattern. The closed-flux horizontal biasing idea is to create, as far as possible, such a divergence-free magnetization pattern in the MRE.

The closed-flux idea is particularly useful in multichannel MRHs where there is not sufficient intertrack spacing to apply the other horizontal biasing schemes discussed earlier in the chapter.

MRH Initialization Procedures

When the fabrication processes of an MRH have been completed, the device is usually subjected to initialization procedures. These procedures can be applied either to the tens of thousands of batch-fabricated heads at the wafer level or to individual finished heads.

The purpose of the initialization processes is to set the proper magnetic state of the various parts of the MRH. Thereafter, the finished MRH has to remain in a reproducible stable magnetic state and be stable against all reasonable magnetic, thermal, and mechanical perturbations.

The parts which typically need to be initialized include those which produce the vertical and horizontal biasing and the MRE. Initialization processes start with those that require high magnetic fields or high temperatures and finish with low-field, low-temperature steps.

When the device contains permanent magnetic materials, these are the first to be initialized. If the PM film has a coercivity, H_c, it is necessary to apply total fields of, typically, 3 to $4H_c$ in order to set the PM film at its maximum remanence. Here, *total field* refers to the vector sum of the external field and any demagnetizing or fringing fields generated by the magnetic layers in the MRH. Very often, external fields exceeding 10,000 Oe are necessary even with PM films that have an H_c of only 1000 Oe. This step sets the remanent magnetization of the hard film at the correct magnitude in the correct direction.

The next process, for devices which contain antiferromagnetic layers used for providing exchange bias, is that of setting the antiferromagnet. Just as ferromagnetic and ferrimagnetic materials have magnetocrystalline anisotropy, with easy and hard axes, so do antiferromagnetic materials. Setting the antiferromagnet means setting the (alternately opposite) spin directions on the correct easy axis of the antiferromagnet.

To initialize an antiferromagnet film, it must be cooled down through its Néel temperature in the presence of a magnetic field. The Néel temperature of an antiferromagnet is the analog of the Curie temperature of a ferromagnet. It is that temperature at which thermal energy ($\frac{1}{2}kT$) destroys the exchange-mediated antiferromagnetic ordering and the material becomes a mere paramagnet. The Néel temperatures of many of the antiferromagnets used in MRH technology are in the range of 300 to 400°C.

Lower temperatures may be used provided longer times are used for the antiferromagnetic film initialization. The blocking temperature is such that the antiferromagnet's anisotropy can be reoriented in a time scale of several seconds; it usually is about 10–20°C below the Néel temperature. The blocking temperatures of nine antiferromagnets commonly considered for use in MRHs are listed in Table 11.1 in Chapter 11. The so-called switch temperature is even lower, at about 20–40° C below the Néel temperature, and is the temperature at which the anisotropy axes can be rotated by 180° in a period of 30 minutes.

The magnetic field, which must be present during the cooling phase, need not be of large magnitude because, at the Néel temperature, the magneto-crystalline anisotropy energy, K, of the material vanishes. The field serves only the purpose of providing a preferred direction during the cooling. It is a magnetic process closely related to that of anhysteresis. Typically, field magnitudes of several hundred oersted suffice and, of course, such low fields do not upset the PM film initialization.

Finally, the magnetic state of the MRE has to be initialized. This process can have two parts. First, it is often necessary to perform a magnetic anneal in order to reestablish the easy axis of the MRE in the correct direction. The correct direction is usually the cross-track direction in both the usual anisotropic MRHs and in the giant MRHs discussed in Chapter 10.

Magnetic annealling in permalloy typically requires several hours at 250°C in a magnetic field. A total magnetic field strength sufficient to saturate the magnetization is required. Higher fields are not more effective. The field strength required is approximately 3 to $4H_c$ plus any demagnetizing and fringing fields that exist. Typically, a field of several hundred oersted suffices. Provided the annealing temperature is less than the switch temperature of any antiferromagnetic material in the structure, the exchange bias initialization is not perturbed.

The development of an induced anisotropy and easy axis in permalloy is called a *magnetization-induced* process and occurs by the development of an ordered atomic structure in which the 81Ni/19Fe spatial distribution of atoms is no longer statistically random, but favors oriented chains of Fe atoms.

The second MRE initialization process is often called "ringing down." The device is exposed to an alternating polarity sequence of magnetic fields which have decreasing magnitudes. The process is entirely analogous to the familiar process of "ac demagnetization," which is used to demagnetize everything from watches and recorded magnetic media to ships. Its purpose is to leave the MRE in the lowest attainable energy state, which, almost by definition, is most likely to have the greatest stability. The initial field magnitudes need only be sufficient to saturate the MRE magnetization. The process is usually performed at room temperature and, accordingly, it does not upset any of the previous initializations.

At the completion of these sequential processes, the MRH is expected to remain in the proper magnetic state throughout its operational lifetime.

Further Reading

Additional reading material is listed here that will prove helpful to readers who seek more detailed information.

Bajorek, C. H., Krongelb, S., Romankiw, L. T., and Thompson, D. A. (1974), "An Integrated Magnetoresistive Read, Inductive Write High Density Recording Head," in *20th Annual AIP Conf Proc.*, American Institute of Physics, p. 24.

Kuijk, K. E., van Gestel, W. J., and Gorter, F. W. (1975), "The Barber Pole, A Linear Magnetoresistive Head," *IEEE Trans.* **MAG-11**, 5.

Markham, David, and Jeffers, Fred (1991), "Magnetoresistive Head Technology," *Proc. Electrochemical Soc.* **90**, 8.

Shelledy, F. B., and Nix, J. L. (1992), "Magnetoresistive Heads for Magnetic Tape and Disk Recording," *IEEE Trans.* **MAG-28**, 5.

CHAPTER

8

Hunt's Unshielded Horizontal and Vertical Anisotropic Magneto-Resistive Heads

As stated in the preface, anisotropic magneto-resistive heads were invented by Robert Hunt at Ampex in 1968. The first paper, written by Hunt, appeared in 1970. Hunt's paper deserves close scrutiny by all interested in MRHs because the principal advantages of using AMRHs and, indeed, GMRHs, rather than the usual inductive heads, are put so clearly. In the first paragraph, Hunt states:

> The following advantages accrue to a magnetoresistive transducer (MRT). 1) The device measures the field rather than the time derivative of flux, making output levels insensitive to scan velocity. 2) The device has no intrinsic frequency limitations for practical bandwidths (i.e., <100 MHz). 3) The device has no intrinsic noise mechanisms to deteriorate signal-to-noise ratio. 4) Many device elements may be mounted side by side on the same substrate and registered to within optical tolerances, eliminating the serious problem of gap scatter in ring head technology. 5) The transducer output levels are characteristically in the 10–100 mV (rms) range. 6) The MRT's wavelength characteristics, except for a gap interference term, are similar to ring head response. A disadvantage of the device lies in its inability to record signals onto tape.

Point 1, that the output voltage is independent of scan velocity, is the principal reason for the widespread application of MRHs in digital tape and small-diameter hard-disk drives. This fact is expanded upon in Chapter 16.

Point 2 is as true today as it was in 1970. Magnetization reversal processes which involve only rotation of the magnetization in a single domain are several orders of magnitude faster than those that proceed by domain wall motion. It is indeed notable that device operation at 100 MHz, capable of supporting some 200 Mbits/second, was being investigated experimentally at Ampex as early as 1968. Such data rates, though of great interest in digital video recording, were not attained in computer peripheral recorders until 25 years later.

Today, point 3 is understood to mean that, when appropriate horizontal biasing is applied in order to suppress domain wall jumps in the MRE end zones, the MRH noise power is simply the Johnson, or thermal, noise ($4kTR\Delta f$) of the MRE. Here, k is Boltzmann's constant, T is the absolute temperature, R is the resistance of the MRE, and Δf is the noise measurement bandwidth. In fact, when the MRE becomes very small, of the order 0.1 μm × 0.1 μm × 20 Å thermal fluctuations cause an additional noise as is discussed in Chapter 19.

Point 4 is precisely the attribute of thin-film MRHs that was exploited in the 0.5-in. tape IBM digital recorders (IBM 3480 and 3490), which have 18 and 36 parallel tracks, respectively. Similarly, the Philips DCC compressed digital audio consumer machines have 16 parallel tracks across a 0.15-in.-wide (4-mm-wide) tape.

Point 5, that output voltages as high as 100 mV rms (over a 50-mil track width) can be realized, is within a factor of 2 of the voltages attained today in permalloy anisotropic MRH heads. We will see in Chapter 17 that the same specific output, that is, the voltage per unit track width, should always be expected from *any correctly designed* anisotropic MRH with a permalloy sensor when it is operated at the same measuring or sensing current density. It is a property of the permalloy only.

Point 6, concerning the MRT's wavelength characteristics or spatial frequency spectrum, is discussed later in this chapter.

Finally, Hunt points out that MRHs cannot record signals, and this is, of course, absolutely correct. It follows that a separate write head must be used with an MRT or MRH. Recall, however, that the physical processes of recording and reading are entirely different, as was reviewed in Chapters 2 and 3. For example, the optimum writing gap is invariably longer than that needed for optimum reading.

When a separate write head is used, however, it becomes possible to accomplish several other desirable changes. For example, the number of turns on the dedicated write head can be reduced, because that number is usually set by reading criteria in dual-role inductive write–read heads. Reducing the number of turns in write heads not only simplifies the head fabrication process, but also reduces the voltage compliance required of the write current driver because the head impedance is reduced. Moreover, the separation of writing and reading heads into separate structures makes possible the strategy of "write wide–read narrow" in which the read channel signal (and signal-to-noise ratio) is deliberately compromised, in order to alleviate track-following problems during read operations.

We can conclude that although Hunt was absolutely correct about the disadvantage of an MRT or MRH, it is now realized that many advantages accrue with the separation of write and read heads mandated by the use of MRHs.

The Unshielded Horizontal Head

Hunt's first MRH is shown in Fig. 8.1. Here the permalloy sensor is laid flat across the recording tape. The MRE senses the medium fringing field component that is in the plane of the thin film. This is the horizontal or x component of the fringing field of the medium. The MRE dimensions used were $W = 50$ mils, $D = 1$ mil, and $T = 1000$ Å.

The horizontal field, measured at a distance y above a tape sinusoidally magnetized uniformly throughout its depth, is known. Let the magnetization waveform $M(x)$ recorded be

$$M(x) = M_R e^{jkx}, \tag{8.1}$$

where M_R is the maximum remanence of the tape or disk, $j = \sqrt{-1}$, k is the wavenumber ($=2\pi/\lambda$), and λ is the wavelength. The exponential term e^{jkx} is called the *generalized sinusoid* and is equal to $(\cos kx + j \sin kx)$.

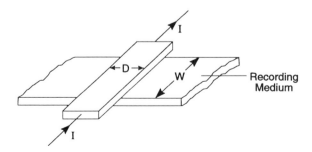

Fig. 8.1. Hunt's horizontal MRH.

The Unshielded Horizontal Head

The horizontal component of the fringing field above the tape is

$$H_x(x) = -2\pi M_R (1 - e^{-k\delta}) e^{-ky} e^{jkx}, \tag{8.2}$$

where δ is the tape thickness and y is the distance above the tape.

The term in parentheses is called the *thickness loss* and shows that 90% of the maximum possible field H_x comes from a surface layer in the medium only about one-third (0.36, actually) of a wavelength deep.

The exponential term e^{-ky} is called the *spacing loss*, and it shows that the field H_x falls to $e^{-2\pi}$ (1/535th) of its value on the surface of the medium when $y = \lambda$. In decibels, the spacing loss is $-54.6y/\lambda$ dB.

We can see from Eq. (8.2) that the magnitude of the horizontal H_x field changes across the element dimension D. Accordingly, the magnetoresistance changes across the dimension D. A simple way to visualize the effect of these changes is to imagine the MRE divided into a number N of strips parallel to the W dimension and the current flow and to consider these strips as a set of variable resistors connected in parallel. Provided the MR coefficient $\Delta\rho/\rho$ is small, it can be shown by Ohm's law that, if each strip has resistance $R + \delta R$, the overall change in resistance of the parallel set is just the average of the changes, $(1/N)\Sigma\delta R$. Of course, the MRE is continuous and the overall change in resistance is actually given by $1/D \int \delta R(x)\, dx$. Here $\delta R(x) = H_x(x)\, dR/dH$, where dR/dH is the slope of the variable resistance versus field characteristic.

Following the straight-line change in resistance versus field approximation of the biased MRE characteristic that was discussed in Chapter 4, the sensitivity or slope is $\frac{1}{2}\Delta R/H_{bias}$. The change in resistance of a single strip is, accordingly, proportional to $\frac{1}{2}\Delta R(H_x/H_{bias})$.

When these changes of resistance are averaged over the MRE dimension D and multiplied by the measuring current I, the result is the output voltage spectrum for small signals ($H_x < H_{bias}$),

$$V_H(k) = \left(\frac{I\Delta R}{2}\right)\left[\frac{2\pi M_R (1 - e^{-k\delta}) e^{-kd}}{H_{bias}}\right]\left(\frac{\sin kD/2}{kD/2}\right), \tag{8.3}$$

where d is the MRE-to-tape spacing.

In this expression, the first bracketed term on the right side, the measuring current times one-half the maximum change in MRE resistance, is obviously a voltage. The other bracketed terms are dimensionless.

The second term is the magnitude of the tape field, H_x, divided by the bias field, H_{bias}. Again, it shows the thickness loss and spacing loss terms common to all read heads that sense the fringing field or flux above a recorded medium.

The third bracketed term is a direct result of the averaging over the distance D discussed earlier. It is precisely the "gap interference" term mentioned by Hunt.

Note that Eq. (8.3) expresses the output voltage spectrum. The spectrum is the magnitude of the output signal at a particular wavenumber, $k(=2\pi/\lambda)$. The actual output voltage waveform is, of course, $V_H(x) = V_H(k)e^{jkx}$. Because the MRH response has been made linear, by using the proper bias field H_{bias}, the output for an input generalized sinusoid at a single spatial frequency k contains only that fundamental frequency. Non-linearities introduce higher harmonics, $2k$, $3k$, etc.

Hunt realized that his horizontal head was not satisfactory because the "gap interference" term severely limits the short-wavelength response. For example, at a wavelength $\lambda = D$, the response is zero. This behavior is analogous to that of an inductive head of gap length g, where the first gap null occurs at $\lambda = g$. The solution to this problem is to turn the MRE on end so that now the "gap interference" term becomes (sin $kT/2)/kT/2$. Since T is the thin-film thickness, the first "gap interference" null now occurs at inaccessibly short wavelengths.

The Unshielded Vertical Head

Hunt's vertical head is shown in Fig. 8.2. Again, it senses only the component of the medium fringing field that is in the plane of the MRE. This is now the vertical, or y, component of the field.

The vertical field, H_y, of the sinusoidally written medium is

$$H_y(x) = j2\pi M_R(1 - e^{-k\delta})e^{-ky}e^{jkx}. \tag{8.4}$$

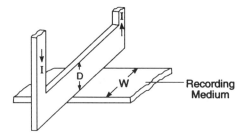

Fig. 8.2. Hunt's vertical MRH.

The Unshielded Vertical Head

The appearance of an extra factor, $j(=\sqrt{-1})$, missing in the analogous expression for the horizontal field H_x, should be noted carefully. It indicates that the vertical field H_y lags 90° out of phase with the written magnetization. If a sine wave is written, the vertical field follows a cosine wave because $\sin(\theta + 90°) = \cos\theta$.

This 90° phase shift is of great importance because a normal inductive reading head also displays a 90° phase shift due to the Faraday law differentiation ($V = -N\, d\phi/dt$) of the head flux. It follows that the digital pulse shape and symmetry of a vertical MRH must be similar and identical, respectively, to that of a normal inductive head.

Different depths of the vertical MRH sense different magnitudes of vertical field H_y. Accordingly, the differing resistance changes over the depth have to be averaged.

The output spectrum for the vertical head is, for small signals ($H_x < H_{\text{bias}}$),

$$V_v(k) = j\left(\frac{I\Delta R}{2}\right)\left[\frac{2\pi M_R(1 - e^{-k\delta})e^{-kd}}{H_{\text{bias}}}\right]\left(\frac{1 - e^{-kD}}{kD}\right)\left(\frac{\sin kT/2}{kT/2}\right). \tag{8.5}$$

The significance of the factor j and the first two terms on the right side has already been discussed. The third term is the result of the averaging of the small resistance changes δR over the depth D of the MRE. At long wavelengths, this term is approximately equal to unity and is thus not an important factor. However, at short wavelengths, $\lambda \leq D$, the term becomes approximately equal to $1/kD$. Thus, the vertical head output voltage falls off at short wavelengths at 6 dB/octave (20 dB/decade).

A simplistic way to think about this severe short-wavelength loss is to realize that the vertical field H_y, sensed by the parts of the MRE most distant from the medium, is very small because it falls off as e^{-ky}. The distant regions of the MRE thus act merely to "shunt out" electrically the resistance changes in the parts of the MRE that are close to the medium.

Despite this severe short-wavelength loss mechanism, in many applications, the extremely high output levels and the simplicity of construction of the vertical MRHs, compared with that of inductive heads, make them the preferred transducer. Vertical heads, which have more recently been called simply unshielded magneto-resistive heads or UMRHs, are used in many bank and credit card readers.

Single Magnetization Transition Pulse Shapes

When considering the performance of both MRHs and inductive heads for applications in digital (or binary) magnetic recording, it is often preferred to consider the so-called "isolated" output pulse shape rather than the output voltage spectrum.

Provided the response of the reproduce head can be considered linear, the pulse shape and the spectrum are, of course, closely related by simple Fourier transforms.

Suppose the written magnetization transition has the arctangent form $M(x) = 2M_R/\pi \tan^{-1}(x/f)$, which was discussed in Chapter 2. The Fourier transform of this magnetization transition, $M(k)$, is equal to $2M_R e^{-kf}/jk$ and the output pulse spectrum is simply $M(k)V(k)$. The output pulse itself is merely the inverse Fourier transform of $M(k)V(k)$.

The advantages of frequency-domain analysis are twofold. First, sequential physical operations result in the simple multiplication of frequency-dependent terms, rather than a tiresome sequence of convolution integrals within convolution integrals. Second, in the frequency domain, the effects of the physical variables have been "separated." Thus, the spectra discussed earlier are the product of terms involving just the medium thickness δ, just the head-to-medium spacing d, just the MRE depth D, and just the element thickness T. A similar separation of variables does not occur when a linear system is described in the spatial or x domain.

Nevertheless, it is instructive to compare the isolated pulse shapes obtained with the vertical MRH and an inductive head.

Consider an inductive head with an almost "zero gap length," which means the gap length g is small compared with the other dimensions (f, δ, d) of the problem.

The isolated output pulse shape is

$$V_R(x) \propto \log\frac{(d+\delta+f)^2 + x^2}{(d+f)^2 + x^2}. \tag{8.6}$$

In the mathematical limit of thin media ($\delta \to 0$), this reduces to the familar Lorentzian pulse shape, $2\delta(d+f)/[(d+f)^2 + x^2]$, so frequently assumed in signal processing studies.

The 50% pulse amplitude width, PW$_{50}$, with the "zero-gap" inductive head is

$$\text{PW}_{50} = 2[(d+f)(d+\delta+f)]^{1/2}, \tag{8.7}$$

which, in the limit of $f \to 0$ and $\delta \to 0$, is just PW$_{50}$ = $2d$. Thus, the smallest imaginable PW$_{50}$ for an inductive head is just twice the read head-to-medium spacing d. This is an interesting piece of knowledge in its own right.

For the unshielded vertical head, a similar analysis for the sharp transition, thin medium case yields

$$V_v(x) \propto \log \frac{(d+D)^2 + x^2}{d^2 + x^2} \tag{8.8}$$

$$_v\text{PW}_{50} = 2[d(d+D)]^{1/2}. \tag{8.9}$$

The preceding expression for the pulse width shows very clearly the undesirable effect of the electrical "shunting" or resistance change averaging over the MRE element depth D in the unshielded vertical MRH. The greater the MRE depth D, the greater the PW$_{50}$. Conversely, note that if D, the MRE depth, could be made negligible ($D < d$), then the limiting pulse width becomes $2d$, just as in the inductive head case.

The solution of this fundamental magnetization transition resolution or isolated pulse width problem in vertical MRHs is discussed in the next chapter.

Side-Reading Asymmetry

Consider the response of an MRE to the off-track magnetic poles shown in Fig. 8.3. When these poles are off-track to the left, the fringing field H_1 is almost orthogonal to the MRE magnetization. Accordingly, H_1 is able to rotate the magnetization by a relatively large angle, because the torque $M \times H_1$ is large.

On the other hand, the fringing field H_2, from magnetic poles off-track to the right, is almost antiparallel to the MRE magnetization. Now the torque is almost zero and the MRE's response is relatively small.

Side-reading, or off-track asymmetry, is inherent in single-element, vertically biased AMRHs. It can be reduced, but not eliminated, by using the barber-pole biasing technique. Perfect off-track symmetry can be

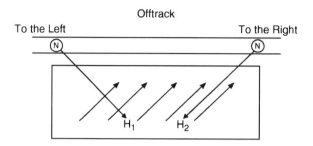

Fig. 8.3. The off-track, or side-reading, asymmetry of a single-element MRH.

achieved in some double-element MRH designs as is discussed in Chapter 15 and in the usual GMR spin valve design discussed in Chapter 10.

Further Reading

Additional reading material is listed here that will prove helpful to readers who seek more detailed information.

Hunt, Robert P. (1970), "A Magneto-resistive Readout Transducer" (digest only), *IEEE Trans.* **MAG-6**, 3.

Hunt, Robert P. (1971), "A Magneto-resistive Readout Transducer," *IEEE Trans.* **MAG-7**, 1.

CHAPTER

9

Single-Element Shielded Vertical Magneto-Resistive Heads

The shielded vertical MRH design overcomes the short-wavelength "shunting" deficiencies of the unshielded AMRH, that was discussed in the previous chapter. The basic design, shown in Fig. 9.1, is simply that of placing a vertical anisotropic magneto-resistive element or GMR spin valve sensor film in the gap between two shields made of a magnetic material of high permeability. To avoid confusion, the space between the MRE and a shield is referred to as the *half-gap* throughout this book.

In Fig. 9.1, only the MRE or sensor film is shown. Note, however, that an actual head will include the extra structures necessary to provide both the proper vertical and horizontal bias discussed in Chapters 6 and 7.

The Function of the Shields

The function of the high-permeability shields can be understood by comparing Figs. 9.2 and 9.3, which show an unshielded and shielded vertical MRH, respectively. In the unshielded configuration, the vertical component, H_y, of the tape or disk fringing field can be large even when the sources of the fringing field, the $\rho = -\nabla \cdot M$ magnetic poles, are far away from the MRE. This "distant sensing" of the unshielded MRH causes its undesirable wide output pulse and short-wavelength spectral roll-off behavior.

9. Single-Element Shielded Vertical Magneto-Resistive Heads

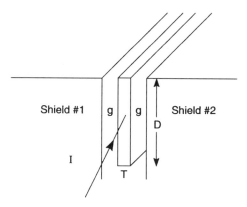

Fig. 9.1. Perspective diagram of a shielded, single-element MRH.

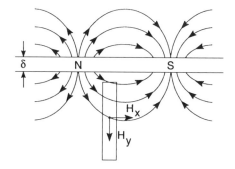

Fig. 9.2. The field and flux flow around an unshielded vertical MRH.

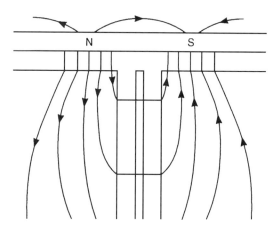

Fig. 9.3. The flux flow through the MRE and shields, when the magnetic transitions are not over the gap.

The Function of the Shields

When shields are added, however, the magnetic flux flow pattern around the recorded medium is changed profoundly. The changes are due to the changed boundary conditions imposed by the shields. On the surface of any magnetic material, the precise boundary conditions are

$$B_{normal}(\text{outside}) = B_{normal}(\text{inside}), \quad (9.1A)$$

$$H_{tangential}(\text{outside}) = H_{tangential}(\text{inside}). \quad (9.1B)$$

When the shield has a very high permeability, the second condition becomes simply $H_{tangential}$ is equal to zero. As shown in Fig. 9.3, the flux from the recorded medium is attracted to the shields (the "keeper" effect) and it has to enter the top of the shields at right angles.

By making the shields extend well below the MRE depth D, very little of the media flux flows through the MRE, as is shown in Fig. 9.3. In this way, the MRE is unresponsive to the effects of distant poles. Moreover, any flux induced in the MRE passes through it normally, causing no change in the angle of magnetization.

When, however, the tape or disk is positioned so that a magnetic pole-bearing region is almost directly over the MRE, the flux pattern changes to that shown in Fig. 9.4. Here magnetic flux from the north pole shown

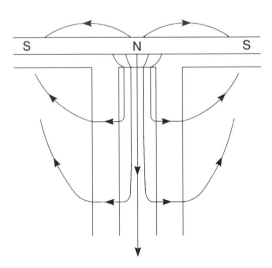

Fig. 9.4. The flux flow through the MRE and shields, when a magnetic transition is over the gap.

flows directly into the top of the MRE. As the flux flows down the MRE, it leaks off sideways through the half-gaps into the shields and returns to the distant south poles. The MRE and shield system is the magnetic analog of an electrical transmission line.

The overall effect of the shields is, therefore, to defer appreciable flux flowing up or down in the MRE until the last possible moment as the poles actually cross the top of the MRE. Recalling that it is the "line-of-sight" magnetic poles that, to first order, always produce the greatest magnetic fields and fluxes, a simple analogy of the shielded MRH's behavior can be made. A pedestrian is only able to see an overflying aircraft as it actually passes directly over an urban canyon, such as Wall Street in New York City, created by adjacent skyscrapers. Similarly, the MRE only "sees" the medium's magnetic poles as they actually pass over the top of the MRE.

To calculate the output voltage spectrum of the shielded MRH, it is necessary to understand not only the magnetic transmission line phenomenon but also the dependence of the MRE flux on the recorded wavenumber k.

The MRE Shield Magnetic Transmission Line

Just as electric charge or current leaks from the center conductor to the outer shield of a coaxial cable, the magnetic flux entering the top of the MRE leaks into the shields.

Consider first the case where the MRE depth is very large as depicted in Fig. 9.5. It can be shown that the flux, $\phi(y)$, flowing down the MRE is

$$\phi(y) = \phi(0)e^{-y/l}, \tag{9.2}$$

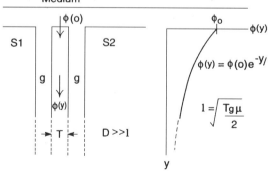

Fig. 9.5. The magnetic transmission line behavior of the MRE and shields.

where $\phi(0)$ is the flux entering the top of the MRE and l is the transmission line decay length.

The decay length is the distance required for the side leakage to reduce the MRE flux to $1/e$ (0.367) of $\phi(0)$. The decay length is

$$l = \left(\frac{Tg\mu}{2}\right)^{1/2}, \tag{9.3}$$

where μ is the permeability of the MRE. The lower the MRE magnetic reluctance (larger T and μ), the longer the decay length. The higher the side-leakage reluctance (larger g), the longer the decay length. The factor of 2 arises because the side leakage occurs into both shields.

Now consider the case where MRE depth D is less than or comparable to the decay length l. This case is shown in Fig. 9.6. Now the MRE flux must become zero at the lower end of the MRE. The situation is analogous to that of an open-circuited electric transmission line, where the open circuit is usually treated by supposing that it is the source of a 180° out-of-phase current wave. Recalling that current in an electric circuit corresponds to flux in the equivalent magnetic circuit, we see that the lower end of the MRE can be considered to be the source of a negative (i.e., upward-flowing) flux.

The net result, shown in Fig. 9.6, is that the MRE flux, $\phi(y)$, decreases almost linearly from the top to the bottom of the MRE. The decrease is at an almost constant slope, when $D \leq l$, because neither the downgoing nor the upgoing flux depart much from constant slope.

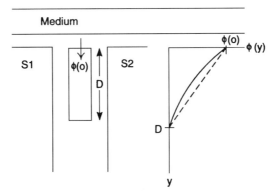

Fig. 9.6. The linear decay of flux with depth when the MRE depth is comparable to the magnetic transmission line characteristic length.

In Chapter 8, we saw that the overall change in resistance of an MRE that is subject to a nonuniform field or flux is proportional to the average flux in the MRE. In the shielded MRH with $D \le l$, the average is obviously always close to 0.5 $\phi(0)$ regardless of the actual depth D.

This observation is important in three ways. First, it shows that the effective efficiency of the shielded MRH is close to 50%. The efficiency is defined in a similar way to that in which the efficiency of inductive read heads is defined. The inductive read-head efficiency is that fraction of the flux entering the top of the head that actually threads the coil.

Second, the fabrication of shielded MRHs is facilitated because the dimension D can have wide tolerances.

Of greatest importance, however, is that, because the MRE thickness T must be chosen so that the top of the MRE does not become magnetically saturated [$\phi(0) \le B_S TW$], all optimally designed single permalloy element shielded MRHs produce approximately *the same maximum specific output* voltage when operated at the same measuring current density.

In some shielded MRH designs, the MRE is not the only highly permeable film between the shields. For example, when the soft adjacent layer (SAL) vertical bias technique is used, the second film has almost the same thickness–permeability product as the MRE. In such designs, because approximately 50% of the flux from the medium flows in the MRE and 50% in the SAL, the optimum design requires that thinner films be used. To keep the top of the MRE and the SAL just below magnetic saturation, their thicknesses are usually reduced to about one-half. It follows that decay the length l is almost unchanged. Moreover, the efficiency of the MRE remains at 50% and, therefore, the maximum specific output voltage is unchanged.

If the MRE is not electrically isolated but is shunted by, for example, a current bias or soft adjacent layer, the maximum specific output voltage is reduced accordingly. When the MRE and parallel shunting layer resistances are R_M and R_S, respectively, the output voltage is reduced by the factor $R_S/R_S + R_H$.

This behavior of the flux $\phi(y)$ in the MRE contrasts sharply with its behavior in the unshielded vertical MRH. The unshielded MRH, of course, does not have the massive shields of high-permeability material that alter the fringing flux pattern of the medium. Because the MRE is so thin (small T), we can assume that it has almost no effect on the medium's fringing flux pattern. Accordingly, the flux flowing down the

Shielded MRH Output Voltage Spectra

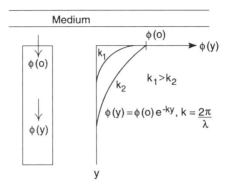

Fig. 9.7. The wavenumber-dependent exponential decay of flux with depth in an unshielded MRE.

MRE, $\phi(y)$, is

$$\phi(y) = \phi(0)e^{-ky}. \tag{9.4}$$

As indicated in Fig. 9.7, the fall-off of flux now depends on the recorded wavelength. The fall-off is just the spacing loss discussed earlier in this book.

The essential difference between an unshielded and a shielded MRH is that the decay of flux in the latter is independent of the recorded wavelength.

Shielded MRH Output Voltage Spectra

The spectrum of the flux, $\phi(0)$, entering the top of the MRE can be elucidated in an almost nonmathmatical manner by an extremely elegant application of the reciprocity theorem. The reciprocity theorem asserts that all (linear) reading heads, which, when energized with current, produce exactly the same fringing field in the tape or disk, must produce exactly the same output flux and voltage spectra and pulse shapes.

Consider Fig. 9.8, which shows the flux flow pattern caused by current flowing in a small coil wrapped around the MRE of a shielded MRH. This coil does not exist in real shielded MRHs. It appears in Fig. 9.8 solely for the purposes of the reciprocity argument. Above the head, the fringing flux flows to the left and right over the left and right half-gaps, respectively. This flux pattern can be decomposed into two parts as indicated in Fig. 9.9.

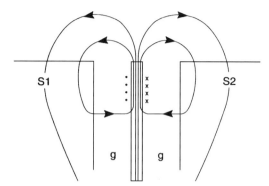

Fig. 9.8. The reciprocity field of a shielded MRH excited by a small coil wrapped around the MRE.

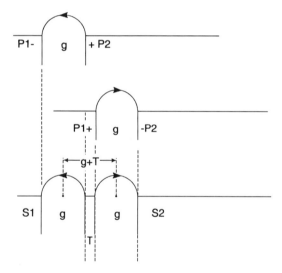

Fig. 9.9. The decomposition of the reciprocity field into the fields of two, oppositely polarized, and horizontally displaced ring heads.

The fringing flux pattern above the left half-gap is almost the same as that above the gap of a normal ring head (top) polarized as shown. Similarly, the flux above the right half-gap is like that of another ring head (bottom) of opposite polarity.

The fringing flux above a shielded MRH, excited by a small coil, is equivalent to that of two normal ring heads of opposite polarity which are

Shielded MRH Output Voltage Spectra

offset from each other by the distance $(g + T)$. By the reciprocity theorem, we can conclude that the flux entering the top of the MRE, $\phi(0)$, must be equal to the difference in the fluxes of the two offset ring heads. Note that this simple conclusion has been arrived at directly, without recourse to mathematical analysis.

The spectrum of the flux flow around a ring head was discussed in Chapter 3. It is, again,

$$\phi_R(k) = -4\pi M_R W \left(\frac{1 - e^{-k\delta}}{k}\right) e^{-kd} \frac{\sin kg/2}{kg/2}, \qquad (9.5)$$

where g is the ring-head whole gap length. The minus term simply means that the flux is 180° out of phase with the written magnetization pattern in the medium.

The MRE flux, $\phi(0)$, is just the difference in the flux of two ring heads that are offset by $(g + T)$. It is convenient to consider the offset of the two heads to be symmetrically $\pm(g + T)/2$. In the spatial frequency domain, these offsets are equivalent to multiplying by $\exp[\pm jk(g + T)/2]$ since the offsets introduce phase shifts of magnitude $\pm k(g + T)/2$ radians.

Recalling that $(e^{jx} - e^{-jx})/2j$ equals $\sin x$, the MRE flux $\phi(0, k)$ can be written as

$$\phi(0, k) = \phi_R(k) 2j \sin k(g + T)/2. \qquad (9.6)$$

Finally, the small signal $(\phi_R < B_s TW/2)$ output voltage spectrum of a shielded AMRH with $D \leq 1$ can be written, by inspection, as

$$V_{\text{AMRH}}(k) = \frac{1}{2}(I\Delta R) \left[\frac{4\pi M_R (1 - e^{-k\delta})}{k}\right] \left[\frac{1}{B_s T}\right] \left(\frac{\sin \frac{kg}{2}}{\frac{kg}{2}}\right) \left(2j \sin \frac{k(g + T)}{2}\right)$$

$$(9.7)$$

Because of the differing definitions of ΔR in AMRHs and GMRHs, the output voltage spectrum of a shielded GMRH, $V_{\text{GMRH}}(k)$, is an identical expression with the sole change that $\Delta R/2$ now appears, not ΔR as in Eq. (9.7).

The several terms that appear on the right-hand side of these spectra can readily be identified with specific physical phenomena. The first term

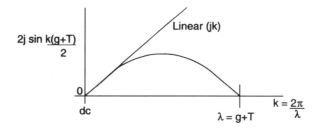

Fig. 9.10. A plot of the ring head displacement function versus wavenumber.

is the 50% reading efficiency of the heads. The second terms are 100% and 50% of the maximum resistance changes in AMR and GMR, respectively. The third term is the flux per unit track width in a small gap-length ring head. The fourth term is equal to the reciprocal of the saturation flux per unit track width of the MRE. The fifth term is the usual Karlquist approximation of the gap-loss function of a ring head with whole gap length g.

The last term, a result of the flux subtraction between the offset ring heads, is plotted in Fig. 9.10. Note the factor j, which shows that the output spectrum suffers a 90° phase change, just as does the output voltage spectrum of an inductive head. At the short wavelength $\lambda = (g + T)$ the shielded MRH has zero output. At medium to short wavelengths, say, $\lambda < 2(g + T)$, the shielded MRH effectively has two "gaplike interference terms." One is due to the half-gap g and the other to their separation $(g + T)$.

For many purposes, a convenient approximation is to pretend that a shielded MRH with half-gap length g has the same spectrum as does a ring head with a whole gap length equal to 1.6 g.

At long wavelengths, the differencing term becomes approximately equal to $2jk(g + T)/2$. The Fourier transform of differentiation, $d(\cdot)/dx$, is jk, and this is also shown in Fig. 9.10. We see, therefore, that the differencing term is an approximation of differentiation. This fact will not surprise those readers who are involved in signal processing, because the standard circuit implementation of differentiation consists merely of subtracting a delayed version of a signal from itself.

The fact that the shielded MRH acts approximately as a differentiator has profound consequences. The normal Faraday's law inductive head output voltage is proportional to $-N\,d\phi/dt$, which is equal to $-NV\,d\phi/dx$, where N and V $(=dx/dt)$ are the number of turns and the head-to-medium relative velocity, respectively. The fact that the shielded MRH output voltage is also proportional to $d\phi/dx$ means that both its spectral shape and pulse output shape closely match those of inductive heads.

The shielded AMRHs and GMR spin valve heads are, therefore, effectively "plug replaceable" with an inductive head. All the coding and preequalization strategies used with inductive heads are equally applicable to shielded MRHs. Concomitantly, all the postequalization and digital bit detection strategies used with inductive heads are appropriate to shielded MRHs. Clearly, this equivalence has greatly facilitated the introduction of shielded MRHs.

Shielded AMRHs were the most widely used MRH design in computer peripheral recorders. They were first used in 0.5-in. IBM tape drives in 1985, IBM small-diameter disk drives in 1991, and 0.25-in. (QIC) tape drives in 1994. In the 0.5-in. drives, the shields are made of NiZn ferrite, which is inherently resistant to mechanical wear. In the small disk and QIC heads, both permalloy and AlFeSil (Sendust) shields are used. It is usual to place the hard, wear-resistant AlFeSil shield on the upstream edge of the MRH. This minimizes the likelihood of the occurrence of gap-smearing damage. GMR spin valve heads for disk drives were introduced by IBM in 1987; all spin valve designs have been shielded.

Shielded MRH Designs

The first shielded MRH designs proposed had the MRE, and its attendant biasing structures, in the gap of a writing head. Thus, the idea was that poles of the writing head, P1 and P2, should double as the shields S1 and S2.

Unfortunately, with this early design, it proved to be almost impossible to maintain the initialization of the MRE. The relatively high deep gap fields, H_o, used in writing heads disturb the desired MRE magnetization state. Deep gap fields are, typically, about equal to three times the coercivity of the recording medium.

The next idea was to fabricate two separate structures. With the writing head poles P1 and P2 entirely separated from the MRH shields S1 and S2, excellent magnetic isolation between the high writing fields and the MRE could be achieved. These structures are referred to as *piggyback* heads.

Later, it was realized that a partial separation of the writing and MRH head was often sufficient and the two heads could share a common pole. In these heads, one of the write-head poles doubles as one of the shields. This technique, shown in Fig. 9.11, is called *merging* and the resultant heads are referred to as *merged heads*.

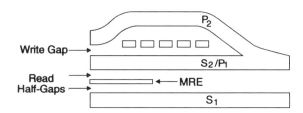

Fig. 9.11. A cross section of a thin-film, merged inductive write head and shielded MRH structure.

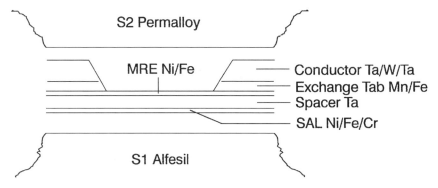

Fig. 9.12. The detailed structure of a SAL-biased, shielded AMRH that uses exchange tabs.

In practice, it is always desirable to deposit the MRE thin films as soon as possible while the structure is still almost flat or planar. Thus, the usual order of deposition of the poles is S1, S2/P1, and P2, as Fig. 9.11 indicates.

In Fig. 9.12, the overall arrangement of the various parts of an IBM 1991-type disk file shielded AMRH is shown. Note the use of exchange tabs for horizontal bias and the use of SAL for vertical bias. The SAL layer is made of NiFeCr, an alloy with a resistivity about a factor of three higher than permalloy, in order to reduce the electrical shunting. The spacer layer is made of high-resistance (tetragonal) phase tantalum. The conductors are composed of low-resistance (cubic) phase tantalum and tungsten layers.

Figure 9.13 shows the configuration of later generation disk file AMRHs. Note that the exchange tabs have been replaced with permanent magnetic hard bias films abutting both the MRE and SAL. In this way, the edge domains of both the MRE and the SAL are controlled.

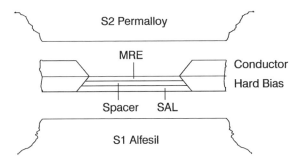

Fig. 9.13. The detailed structure of a SAL-biased, shielded AMRH that uses hard bias for both the MRE and SAL.

The design of spin valve GMR heads is covered in Chapter 10.

Further Reading

Additional reading material is listed here that will prove helpful to readers who seek more detailed information.

Bajorek, C. H., Krongelb, S., Romankiw, L. T., and Thompson, D. A. (1974), "An Integrated Magnetoresistive Read, Inductive Write High Density Recording Head," in *20th Annual AIP Conf. Proc.*, American Institute of Physics.

Potter, Robert I. (1974), "Digital Magnetic Recording Theory," *IEEE Trans.* **MAG-10**, 3.

Zhang Yun, Shtrikman, S., and Bertram, H. (1997) "Playback Pulse Shape and Spectra for Shielded MR Heads," *IEEE Trans.* **MAG-32**, 2.

CHAPTER

10

Simple Spin Valve Giant Magneto-Resistive Heads

The giant magneto-resistive (GMR) effect was not known to science until 1988, but this did not deter MRH designers from incorporating the effect into working prototype GMRHs by 1993 and starting mass production in 1997. Usually, the GMR structure or material has been used as the magneto-resistive element (MRE) in the shielded configuration discussed in Chapter 9. Today, GMR heads (GMRHs) are the only type used in disk files.

The attraction of using GMRHs lies in both their greater sensitivity and their higher total change in resistance. For small-signal analog applications, the GMRH is superior in proportion to its low-field sensitivity, $d(\Delta\rho/\rho_0)dH$. In optimized large-signal digital GMRHs, their advantage is proportional to the magneto-resistive coefficient $\Delta\rho/\rho_0$.

Spin Valve GMRHs

The structure of the prototypical spin valve is shown in Fig. 10.1. There are just two ferromagnetic layers of permalloy. They are called the *pinned* and the *free* layers. The pinned layer is exchange coupled to an underlying antiferromagnet with the requisite intimate atomic contact being indicated schematically in Fig. 10.1 by the interfacial zigzag line.

The atomic details of an idealized, epitaxial ferromagnet/antiferromagnet exchange-coupled interface are shown in Fig. 10.2. If every atom

Spin Valve GMRHs

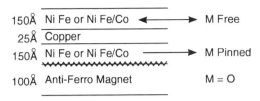

Fig. 10.1. A permalloy–copper spin valve GMR structure showing the exchange pinned and the free layers.

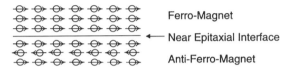

Fig. 10.2. The magnetic state of a nearly perfect, or epitaxial, interface between a ferromagnet and an antiferromagnet.

couples ferromagnetically across the interface, then very high exchange anisotropy can result. If, however, the atoms misregister by, say, one atom in 100, then the exchange coupling will be some average over the varying interatomic spacings and might be low. Alternately, if monoatomic steps exist in the antiferromagnet, the overall coupling will again be low. Furthermore, if the antiferromagnet–ferromagnet interface is such that alternate spins in the antiferromagnet are oppositely oriented, which is the case in the so-called "compensation" crystallographic planes of some antiferromagnets, no coupling will occur. As has been pointed out in Chapter 6, the exact atomic details of the interface are critical and might not be measurable, because they are buried beneath the overlying permalloy layer.

The purpose of exchange coupling the pinned layer to the antiferomagnet is to establish its magnetization in a fixed direction. As discussed in Chapter 7, this direction can be set during the initialization processing of the head. Typically, the antiferromagnetic layer is a few hundred angstroms thick and is made of MnFe, CoO, or MnIr. Various properties of several, potentially useful, exchange coupling antiferromagnetic materials are discussed in Chapter 11.

The pinned layer is usually made of permalloy, for all the usual reasons. Because copper (but not silver) is soluble in permalloy, it is often overcoated ("dusted") with a thin, 5- to 10-Å layer of cobalt, which acts as a diffusion barrier stopping the copper–permalloy interdiffusion.

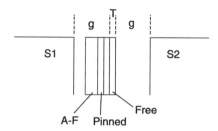

Fig. 10.3. The arrangement of a spin valve GMR structure in the gap of a shielded head.

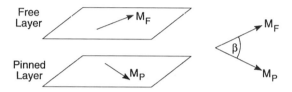

Fig. 10.4. The free and pinned layers of a spin valve showing the angle between the magnetizations.

Above the nonmagnetic layer, usually of copper, is the cobalt-dusted permalloy free layer. In the free layer, the anisotropy axis, defined during deposition or by magnetic annealing during initialization, is usually set in the cross-track direction, that is, orthogonal to the pinned magnetization direction. The reason for this "crossed" arrangement is discussed later.

Figure 10.3 shows how the entire spin valve structure is placed between shields to make a shielded GMRH. Note that if the free layer is not on the center line between the shields, self-biasing occurs.

In Fig. 10.4, just the free and pinned layers are shown in perspective. The angle between their single-domain magnetization directions is β and the GMR is proportional to $-\Delta R/2 \cos \beta$.

The perspective diagram shown as Fig. 10.5 again shows the free and pinned layer magnetization directions and, additionally, the crossed easy axis of the free layer.

The equilibrium magnetization–easy axis angle θ, shown in Fig. 10.6, is, in the absence of demagnetizing field effects, simply given by $\sin \theta = H_y/H_k$, where H_y is the magnetic field from the recording medium and H_k is the anisotropy field. The angle between the free and pinned layer magnetizations, β, is equal to $(\pi/2 - \theta)$. Accordingly, the change in GMR,

Spin Valve GMRHs

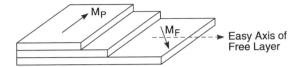

Fig. 10.5. A perspective diagram of a spin valve showing the easy axis of the free layer.

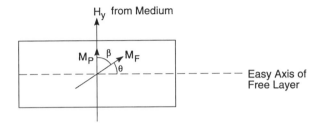

Fig. 10.6. The effect of the recording medium field on the free layer magnetization angle.

δR, is

$$\delta R = -\frac{\Delta R}{2}\cos\beta = -\frac{\Delta R}{2}\cos\left(\frac{\pi}{2} - \theta\right)$$

$$= -\frac{\Delta R}{2}\sin\theta = -\frac{\Delta R}{2}\frac{H_y}{H_k}. \qquad (10.1)$$

The remarkable fact thus emerges that the change in resistance, δR, is directly proportional to the vertical field, or flux, from the recording medium. This linear characteristic is shown in Fig. 10.7. Note that this intrinsically linear characteristic has been obtained without the use of any of the vertical bias techniques that are required in anisotropic MRHs. It is, of course, still necessary to provide the vertical exchange bias on the pinned layer and the appropriate horizontal bias on the free layer in order to keep it single domain.

Naturally, in a real device the finite dimensions of the free layer cause nonuniform demagnetizing fields to exist. These make it increasingly difficult to magnetize the free layer in the vertical (up or down the gap) directions. The effect of the nonuniform demagnetizating fields causes the resistance change characteristic to enter saturation gradually, displaying the wings shown in Fig. 10.7.

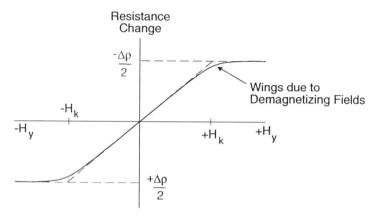

Fig. 10.7. The GMR resistance change versus field characteristic of a spin valve showing the effect of nonuniform demagnetization fields.

For an unshielded spin valve without hard bias fields applied the effective anisotropy field is $H_k + M(N_y - N_z)$ where H_k is the magnetocrystalline anisotropy field, $2K/M$, and N_y and N_z are the vertical and horizontal demagnetizing factors, respectively. In practice, the shape anisotropy fields are much larger than H_k, so that the initial slope of the ΔR versus H_y characteristic is approximately $\Delta R/2M(N_y - N_z)$. When a horizontal bias field of magnitude H_{HARD} is applied, the initial slope is reduced and becomes $\Delta R/(2M(N_y - N_z) + H_{HARD})$.

The inevitable consequence of using the necessary horizontal bias is to reduce the sensitivity of an MRH. Accordingly, the magnitude of horizontal bias has to be limited to that necessary to maintain the stability of the single-domain state against credible stray fields. A common specification is that there be no stray field above 25 Oe at the site of the MRH when it is installed in a disk drive. The typical horizontal field, $H_{HARD} \approx 4\pi MT/W$, is, for a permalloy free layer 50 Å thick and 5000 Å wide, about 100 Oe.

In practice, it is found that the spin valve resistance change characteristic is not of the exact odd symmetry shown in Fig. 10.7. The pinned layer is effectively a permanent magnet and it produces, by magnetostatic coupling, an unwanted vertical bias field on the free layer. This vertical bias field may be cancelled out by operating the spin valve at exactly the correct magnitude (and polarity) current so that the current bias field is equal and opposite to the pinned layer bias field. Note that a correctly balanced spin valve GMRH should have neither pulse amplitude asymmetry nor, to first order, side-reading asymmetry, because the quiescent angle, θ, is zero.

The dynamic range of magnetization angle change, set by the acceptable (odd only) harmonic distortion in a spin valve head, is approximately ±60° about the quiescent θ = 0°. The anisotropic MRH dynamic range is about ±30° about the quiescent 45°. Note, however, that fractional change in the resistance is the same. Thus, $\Delta R/2(\cos 30° - \cos 150°) = 0.87\Delta R$ for the GMR head, whereas $\Delta R(\cos^2 15° - \cos^2 75°) = 0.87\Delta R$ in the anisotropic MRH case.

Following the shielded AMRH analysis given previously in Chapter 9, the output voltage spectrum of a horizontally biased, shielded GMR spin valve can be expressed immediately as

$$V_{\text{GMRH}}(k) = \frac{1}{2}\left(\frac{-I\Delta R}{2}\right)\left[\frac{4\pi M_R(1-e^{-k\delta})e^{kd}}{k}\right]$$

$$\times \left[\frac{1}{B_s T}\right]\left(\frac{\sin\frac{kg}{2}}{\frac{kg}{2}}\right)\left(2j\sin k\left(\frac{g+T}{2}\right)\right) \quad (10.2)$$

Again the individual terms can be identified: the 50% efficiency, the MR sense current times one-half the GMR maximum resistance change, the flux from the disk per unit track width, the reciprocal of the free layer maximum flux content per unit trackwidth, the normal half-gap Karlquist loss factor, the 90° phase shift (leading relative to the written magnetization) and, finally, the subtraction factor between the two half-gaps. Apart from the polarity change, the output spectrum and, therefore, the isolated output pulse shape is the same as for shielded AMRHs and inductive reading heads.

Further Reading

Additional reading material is listed here that will prove helpful to readers who seek more detailed information.

Tsang, C., Fontana, R., Lin, T., Heim, D., Speriosu, V., Gurney, B., and Williams, M. (1994), "Design Fabrication and Testing of Spin-Valve Read Heads for High Density Recording," *IEEE Trans.* **MAG-30**, 6.

CHAPTER
11

Enhanced Spin Valves

The first GMR spin valves, discussed in Chapter 10, had a number of inherent limitations. For example, the antiferromagnet/pinned layer coupling was often not adequate and the resistance characteristic of the head was not of exact odd symmetry. In this chapter, many of the early modifications made to enhance the performance of GMR spin valves are discussed.

Top and Bottom Spin Valves

There are two ways to construct the spin valve, as is shown in Figs. 11.1 and 11.2.

The basic choice is whether to start or finish with the antiferromagnet layer. In the "top" design it is relatively easy to arrange that the "hard" bias act on the free layer properly. On the other hand, critical control over the pinned–AFM layer interface becomes more difficult because the pinned layer overlays several other layers and, of course, the more layers, the less controlled the interface. The "bottom" design, where the AFM is deposited as soon as is possible, minimizes the number of layers below it and is, therefore, more easily controlled. However, in the bottom design arranging the hard bias is more difficult. Most modern spin valves are "bottom" designs, which reflects upon the fact that ensuring proper exchange coupling of the pinned layer is the more difficult challenge.

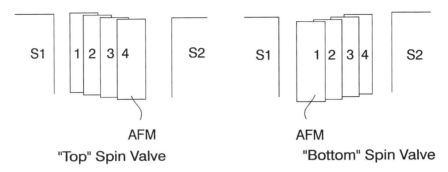

Fig. 11.1. The order of depositions in "top" and "bottom" spin valves.

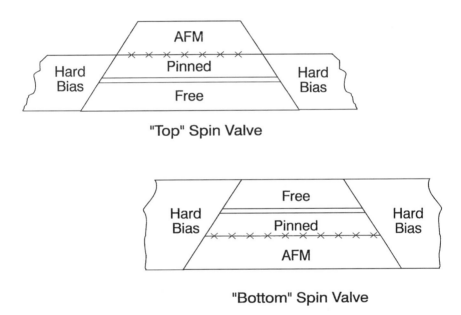

Fig. 11.2. Cross sections of "top" and "bottom" spin valves.

Exchange Coupling Phenomena

A great deal of attention has been devoted to increasing the pinned layer–AFM exchange coupling field. The pinned–AFM geometry is shown in Fig. 11.3 where the exchange coupling in the interface is shown as crosses.

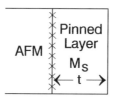

Fig. 11.3. The exchange coupling between an AFM and the pinned layer.

Suppose the exchange coupling energy on the interface is E_{EX} ergs/cm^2. The exchange coupling field, H_{EX}, is then

$$H_{EX} = E_{EX}/M_s t \qquad (11.1)$$

where M_s and t are the saturation magnetization and thickness of the pinned layer, respectively.

It is clear that the thicker the pinned layer, the lower becomes the exchange coupling field, H_{EX}. This causes a dilemma for the designer because, in order to achieve a large GMR effect, the pinned layer thickness must be comparable to the majority carrier penetration length, λ_1, which is about 50–100 Å in permalloy. The exchange field must be greater than the sum of all other magnetic fields; otherwise, control of the magnetization direction in the pinned layer will be lost. This occurrence is called "unpinning" but, unlike the antiferromagnet upset phenomenon discussed in Chapter 13, it is reversible in the sense that when the magnetic fields are removed, the pinned layer becomes again correctly magnetized.

The direction of the exchange field can be set, or changed, by magnetic annealing. In magnetic annealing, an external field is used to hold the magnetization of the pinned layer in the desired direction. A field magnitude sufficient to saturate the pinned layer is sufficient with no benefit from higher fields. This magnetic field has, of course, no direct action upon the antiferromagnet because it has zero magnetization. The only coupling with the AFM is via the interfacial exchange interaction. The process of magnetic annealing is to cool the AFM down through its Néel temperature or blocking temperature while the pinned layer is held magnetized by the external field. At the Néel temperature, the antiferromagnetic ordering of the spins within the AFM is lost because, by definition, the exchange coupling becomes equal to zero. At the blocking

TABLE 11.1
Blocking Temperatures, Exchange Coupling Energies, and Critical Thicknesses of Several Potentially Useful AFMs

Antiferromagnet	Blocking temperature (°C)	E_{EX} (erg/cm^2)	Critical thickness (Å)
$Ni_{45}Mn_{55}$	375	0.24	300
$Pt_{49}Mn_{51}$	340	0.20	300
α-Fe_2O_3	320	0.10	<500
$Pt_{20}Pd_{30}Mn_{50}$	300	0.12	250
$Cr_{40}Mn_{40}Pt_{10}$	280	0.90	300
$Ir_{20}Mn_{80}$	280	0.15	80
$Ru_{12}Rh_8Mn_{80}$	225	0.17	100
$Ni_{50}O_{50}$	210	0.12	400
$Fe_{50}Mn_{50}$	180	0.11	110

temperature, which is usually some 10–20°C below the Néel temperature, thermal fluctuations randomize the directions of antiferromagnetic ordering within individual grains in a manner analogous to superparamagnetism in a ferromagnet. When the AFM grains are large there is little difference between the blocking and Néel temperatures, and conversely with small-grain AFMs.

The magnetic anneal procedure is then:

(a) Heat above the blocking temperature
(b) Apply external field in the chosen direction
(c) Cool to below the blocking temperature
(d) Turn off the external field

In Table 11.1, the blocking temperatures of several antiferromagnets are given together with the interfacial exchange coupling energy, E_{EX}.

It should be noted that, contrary to simplistic expectations, a high value of E_{EX} need not correlate with a high blocking temperature. This is because the actual E_{EX} obtained depends upon the fraction of AFM grains that are correctly oriented to permit interfacial exchange coupling, as is discussed in Chapter 6. That fraction is usually rather low, being in the 5–10% range only.

Table 11.1 also gives a listing of critical thicknesses. This is the thickness of the AFM film that is required in order to realize the listed

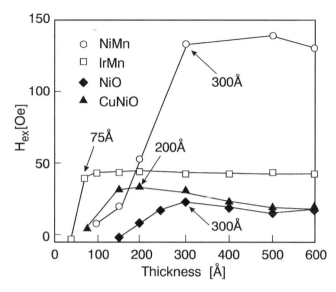

Fig. 11.4. Exchange field, H_{EX}, versus AFM layer thickness for several AFMs coupled to 250 Å permalloy.

values of E_{EX}. The thicknesses vary widely from 80→500 Å and reflect upon the different growth habits of the AFM films. When the sputtering of the AFM film starts the percentage of favorably oriented grains is low and, as the deposition proceeds, it increases and eventually saturates. This phenomenon is shown also in Fig. 11.4, where all the AFMs are coupled to a 250 Å thick permalloy layer. It should be noted that both, it increases and eventually saturates.

Table 11.1 and Fig. 11.4 indicate that IrMn has not only the lowest critical thickness but also a satisfactorily high blocking temperature. A high blocking temperature minimizes the possibility of the spin valve suffering the antiferromagnet upset phenomenon discussed in Chapter 13.

Another aspect of the exchange coupling is displayed in Fig. 11.5, which shows the influence of temperature upon the temporal decay of the exchange field in an IrMn/CoFe pair.

The exchange field, H_{EX}, normalized to unity at zero time, is plotted versus the logarithm of time at various temperatures in the range 100–200°C, while the AFM–ferromagnetic pair is exposed to a constant 3000-oersted field opposing the exchange field. Although the deleterious

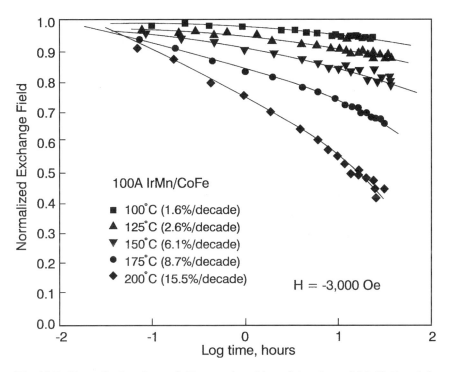

Fig. 11.5. Normalized exchange field versus logarithm of time for an IrMn/CoFe pair in an opposing 3000-oersted field.

effect of the higher temperatures is evident, it appears that at 100°C, the IrMn/CoFe pair is satisfactorily stable with only about 1.6% decrement per decade of time. These decay phenomena are analogous to those occurring in the onset of superparamagnetism found in fine-grained ferromagnets, such as those used in modern thin-film hard disks.

Cobalt "Dusting" and CoFe Alloys

The first spin valves used permalloy (81 Ni/19 Fe) for both the pinned and free layers. When a copper spacer layer was used, it was found that, because copper diffuses into permalloy, the spin valve's performance degraded rapidly over time. In order to prevent this degradation, an antidiffusion barrier was required. Suitable materials for the antidiffusion

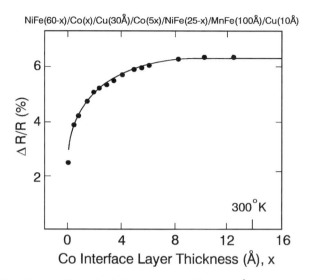

Fig. 11.6. Showing the effect of cobalt nanolayer thickness, x Å, on the GMR coefficient.

barrier are, of course, those that are immiscible in both the AFM and the permalloy. They have either different crystal habits or 10–15% mismatched atom–atom spacings. Cobalt was selected, and it was found that not only did a two or three atom thick (4–6 Å) "dusting" of cobalt stop the interdiffusion, but also that it enhances the GMR coefficient markedly, as is shown in Fig. 11.6. Here the GMR coefficient, $\Delta R/R$, is plotted against the cobalt nanolayer thickness, xÅ in a spin valve with the structure NiFe(60)/Co(x)/Cu(30)/Co($5x$)/NiFe($25x$)/ MnFe(100)/Cu(10). The presence of the cobalt increases the GMR by almost a factor of 3. Notice that one of the cobalt layers is thin, up to 12 Å, but that the cobalt on the pinned layer is five times thicker. As has been discussed in Chapter 5, a precise understanding of the physics underlying these interfacial effects is lacking. A recent trend is to use CoFe alloys for both the pinned and free layers and dispense with the "dusting" altogether.

Spin Valve Biasing Effects

In a simple spin valve the pinned layer is held by exchange coupling to the antiferromagnet, magnetized either up or down. Magnetostatically, it behaves just like a permanent magnet and, accordingly, it produces a

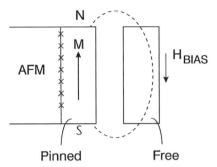

Fig. 11.7. The vertical bias field in the free layer from the pinned layer.

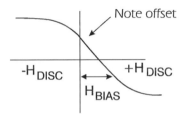

Fig. 11.8. The offset GMR resistance versus disk field characteristic curve.

vertical, or transverse, bias field in the free layer, as is shown in Figs. 11.7 and 11.8.

The offset in the resistance versus field characteristic curve is undesirable. It causes, of course, both pulse amplitude asymmetry and track profile asymmetry. Both are the consequence of the nonzero quiescent magnetization angle in the free layer.

The offset can be corrected by (a) choosing the proper direction for the measuring current and (b) choosing the proper magnitude of the measuring current. As appears in Fig. 11.9, when the pinned layer is held magnetized up, the current direction of the measuring current is out of the plane of the diagram.

The measuring current flows in the pinned, spacer, and free layers in proportion to their resistances. The current flowing in the pinned and spacer layers produces a current bias vertical field in the up direction in the free layer, and it can be adjusted to cancel exactly the down permanent-magnet-like bias field from the pinned layer. It should be noted that this method of balancing the resistance characteristic *requires* the correct

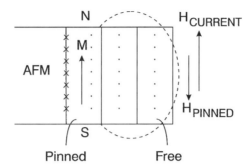

Fig. 11.9. Using current bias to cancel the permanent-magnet-like bias of the pinned layer.

polarity of the measuring current. The GMR spin valve head now has to be operated not only with the correct measuring current polarity but also with the correct direction of head–disk relative motion as was discussed in Chapter 9.

The spacer layer in spin valves is usually made of a "noble" metal such as copper, silver, or gold. With noble metal space layers only low-magnitude RKKY coupling fields exist. Nevertheless, if the spacer layer thickness is chosen correctly, the RKKY coupling field opposes the pinned-layer permanent-magnet-like bias field. No matter whether the pinned layer is held magnetized up or down, when the RKKY coupling is positive, it tries to hold the pinned layer and free layer magnetizations parallel.

Yet another approach to balancing the spin valve resistance characteristic is the so-called bias compensator layer design shown in Fig. 11.10.

In this design an additional layer of Ni/Fe is added and it is separated from the free layer by an insulator layer of thickness equal to that of the noble metal spacer. The bias compensator layer thickness and composition are made the same as in the pinned layer. When measuring current of sufficient magnitude flows into the plane, it is able to saturate the magnetization of the bias compensator layer. Then, by symmetry about the free layer, the pinned layer bias field and the bias compensator layer bias field are equal and opposite and cancel out. There remains the current bias field in the free layer, and this is usually cancelled out by using the self-bias technique. As was described in Chapter 6, self-bias occurs whenever the sensing element is *not* on the centerline between the shields S1 and S2.

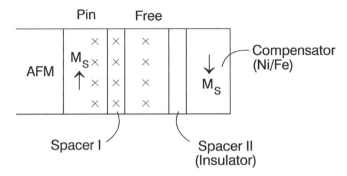

Fig. 11.10. The bias compensator layer design.

In the case shown in Fig. 11.10, an up self-bias field is required, and this means that the element must be displaced to the right-hand side of the center line. In the bias compensator layer design, the spacer between the free layer and the bias compensator layer must be an insulator. If a noble metal spacer were used, the output voltage of the complete structure would be virtually zero because there would be two equal spin valves of opposite polarities connected in parallel.

Unpinning Phenomena

When the measuring current in a simple spin valve is used to balance the resistance characteristic curve, it produces a down vertical current bias field upon the pinned layer that opposes the up exchange coupling field, H_{EX}, between the AFM and the pinned layer. If the current bias field magnitude exceeds the exchange field, the magnetization of the pinned layer will reverse. This phenomenon is called "unpinning," and it is reversible, of course, in the sense that when the down current bias field is removed, the magnetization in the pinned layer will again become upwardly directed.

It should be noted that, in the bias compensator design, unpinning cannot occur because the measuring current direction is reversed. When the measuring current is out of the plane, as indicated in Fig. 11.10, it produces an up vertical bias field in the pinned layer which adds to the exchange bias field.

Further Reading

Additional reading material is listed here that will prove helpful to readers who seek more detailed information.

Gurney, B., Cazey, M., Tsang, C., Williams, M., Dazkin, S., Fontama, R., and Grochowski, E. (1999), "Spin Valve Giant Magnetoresistive Sensor Materials for Hard Disc Drives," *Datatech Mag.*, December.

Lin, T., Mauri, D., and Luo, Y. (2000), "A Ni–Mn Spin Valve for High Density Recording," *IEEE Trans. Mag.* **36**(5), 2563–2565.

CHAPTER

12

Synthetic Antiferro- and Ferrimagnets

This chapter deals with two relatively recent (about 5 years old) developments in the continuing evolution of GMR spin valves. The first is a means of increasing the effective exchange coupling field, H_{EX}, by using a so-called synthetic antiferromagnet in the pinned layer. The second development is a technique for allowing the free layer to be thick, in order to enhance the GMR coefficient, but at the same time to have a low flux capacity, as is required to match properly the recording layer.

In the Pinned Layer

In the previous chapter, it was shown that if the interfacial exchange coupling energy between the AFM and the pinned layer is E_{EX} ergs/cm^2, the exchange field $H_{EX} = E_{EX}/M_s t$ where M_s and t are the saturation magnetization and thickness of the pinned layer, respectively. In order to minimize the possibility of the phenomenon, discussed in Chapter 11, of unpinning occurring and to increase the thermal stability, it is necessary to increase the magnitude of H_{EX}. In order to increase H_{EX} it is clear that either E_{EX} must be increased or the product $M_s t$ must be decreased. Unfortunately, it does not appear likely that E_{EX} can be increased substantially because, even after at least a decade of development, it still remains at only 5–10% of

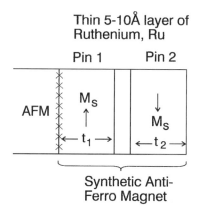

Fig. 12.1. The AFM/synthetic antiferromagnet pinned layer structure.

the maximum theoretical value. The several possible reasons for the low coupling are reviewed in Chapters 6 and 10. The possibility of effecting a substantial reduction in the pinning layer $M_s t$ product is likewise low, because, as is discussed in Chapter 5, high GMR requires a magnetic material of thickness comparable to the majority spin penetration depth, λ_1.

The solution to this design dilemma is shown in Fig. 12.1.

Here the pinned layer structure is made up of three layers. The two outer layers of Co/Fe are separated by an extremely thin layer of ruthenium, Ru, one of the refractory metals. The actual Ru layer thickness, 7 Å, is chosen to coincide with the first minimum of the RKKY exchange coupling curve. The RKKY negative coupling forces the two outer layers of Co/Fe alloy to be oppositely magnetized, in the manner of an antiferromagnet. The RKKY antiferromagnetic coupling fields ($E_{EX}/M_s t_1$ and $E_{EX}/M_s t_2$) are extremely high, being approximately 5000 Oe. The key to understanding the synthetic antiferromanget technology is to realize that since the flux content (per unit width) of the three-layer, tightly bound, pinned structure is now ($M_s t_1 - M_s t_2$), the AFM synthetic pinned layer exchange field,

$$H_{EX} = E_{EX} / (M_s t_1 - M_s t_2), \qquad (12.1)$$

can become very large. In fact, if $M_s t_1 = M_s t_2$ exactly, the H_{EX} would become equal to the RKKY Co/Fe–ruthenium antiferromagnetic exchange field.

In practice, $M_s t_1 \neq M_s t_2$ and the three layer structure should properly be called a synthetic *ferrimagnet*, but this nicety is not always observed.

The first layer, t_1, is exchange coupled to the AFM in the usual manner. The third layer, t_2, acts as the reference or pinned layer in the spin valve.

Synthetic ferrimagnets are now widely employed in advanced GMR spin valves, and they make possible effective AFM–pinned layer coupling fields in the range 500–1000 Oe. This technique permits the use of AFMs which do not produce sufficiently high exchange fields when used in a single-layer pinned spin valve design, but which have other desirable properties such as a high blocking temperature or being an electrical insulator.

In some spin valves, the three-layer synthetic ferrimagnet is used alone without using an AFM. This is possible because the effective anisotropy field, $_{\text{eff}}H_k$, of the tightly coupled ferromagnets is

$$_{\text{eff}}H_k = \frac{2(Kt_1 + Kt_2)}{M_s t_1 - M_s t_2}, \qquad (12.2)$$

where K is the induced magnetocrystalline anisotropy constant of the ferromagnetic layers. Clearly, when $M_s t_1 \approx M_s t_2$, the effective anisotropy field can become high enough to engender stability without requiring the usual AFM.

Other refractory metals, such as rhodium, iridium, rhenium, and platinum also yield extremely high RKKY fields. However, ruthenium has emerged as the interlayer material of choice not only in spin valves but also, acting in precisely the same manner, in the most recent thin-film disks. In this case, the objective is to raise the energy barrier ($KV = \frac{1}{2}H_k MV$) against thermal energy and thus reduce the superparamagnetic decay of the written magnetization.

A thin interlayer of 7 Å is but three atoms in thickness and it is, indeed, a remarkable testimonial to the control possible in thin-film deposition equipment today. The ruthenium layer forms epitaxial interfaces with the Co/Fe because it has the same crystal structure (hccp) and atom–atom spacings and one might, therefore, expect interdiffusion problems. It is remarkable, however, that such a thin layer is stable against interdiffusion, as experiments at elevated temperatures (200°C) for extended periods of time (several weeks) have proven. No doubt the high

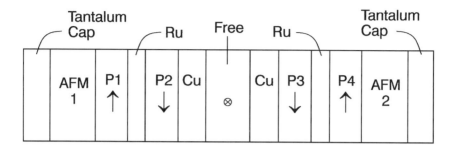

Ta - 25, Ni-MN-150, Co-Fe-30, Ru-8, Co-Fe-25, Cu-20, Ni-Fe-50, Cu-20, Co-Fe-25, Ru-8, Co-Fe-30, Ni-MN-150, Ta -25

13 LAYERS !

Fig. 12.2. A dual synthetic spin valve.

atom–atom binding energy, which is proportional, of course, to the melting temperature, accounts for both this stability and, indeed, the selection of refractory metals for the interlayer.

Another recent development is the so-called dual synthetic spin valve shown in Fig. 12.2.

It will be seen that this structure has no fewer than 13 primary layers and that it contains two synthetic ferrimagnet pinned layers arranged symmetrically about the central free layer. It can be shown that this structure produces a factor 4/3 or 133% higher output than a single spin valve. Note that because layers P1 and P4 are magnetized in the same direction, the magnetic annealing of the AFM1 and AFM2 can be accomplished simultaneously with an external magnetic field. A dual synthetic spin valve should display neither pulse amplitude nor track profile asymmetry because the free layer quiescent magnetization angle is zero.

The dual synthetic spin valve design offers, however, no cancellation or protection against thermal asperities (TAs) because the two spin valves are sensed in addition. Dual *stripe* synthetic spin valves have been investigated, and they have the structure shown in Fig. 12.3. In this design, the central free layer is split into two halves by an insulating film, thus permitting the subtractive or differential that is essential for the cancellation of TAs. Note that now the synthetic ferrimagnet pinned layers must

In the Free Layer

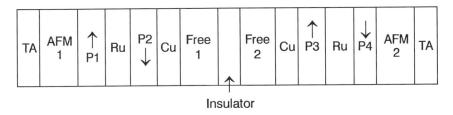

Fig. 12.3. A dual stripe synthetic spin valve.

Ta25, NiMn150, CoFe30, Ru8, CoFe25, Cu20, NiFe50, Al$_2$O$_3$100, NiFe50, Cu20, CoFe25, Ru8, CoFe30, NiMn150, Ta25

15 LAYERS !

be magnetized oppositely with P1 and P3 up and P2 and P4 down, so that the left and right spin valves operate with opposite polarities. With opposite polarity, the differential sensing yields a factor of 2 or 200% of the output of a single spin valve. It follows that the magnetic annealing of the AFMs has to be performed using current bias to provide the magnetic field. The correct annealing field, up on the left and down on the right, is produced by currents of the same polarity in the two sides of the structure.

In the Free Layer

Synthetic ferrimagnets can also be used to advantage in the free layers of GMR spin valves. As is discussed in Chapter 18, the flux capacity, $4\pi M_s TW$, of the free, or sensor, layer has to be matched to the flux capacity, $4\pi M_R \delta W$, of the recording medium. As recording technology progresses to high linear densities, recording media of ever-reduced areal flux capacity, $4\pi M_R \delta$, are being used because, as is reviewed in Chapter 2, lower $M_R \delta$ media yield smaller written slope parameters, f, and this, in turn, allows high linear densities to be used without excessive overlap of the written magnetization transitions. Accordingly, the thickness, T, of the free layer has been and will continue to be reduced, as can be seen, for example, in Table 18.1.

On the other hand, in order to achieve a high GMR effect, the free layer thickness has to be about equal to the majority spin current carrier's penetration, or coherence length. This can be seen very clearly in Fig. 5.9.

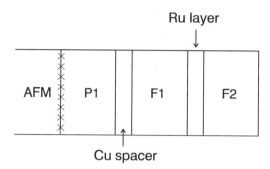

Fig. 12.4. A spin valve with a synthetic ferrimagnet sensor structure.

The solution to this design dilemma is to use the thin-layer, synthetic ferrimagnet sensor structure shown in Fig. 12.4.

Again, the three-layer structure comprises outer relatively thick layers of Co/Fe alloy sandwiching a thin (7 Å) layer of ruthenium. The intense RKKY antiferromagnetic coupling forces the outer layers to be magnetized oppositely. In the quiescent state, these magnetizations will lie in the cross-track direction along the induced anisotropy easy axis of the permalloy-like layers. The effective flux capacity is $4\pi M_s(t_1 - t_2)$ with an effective layer magnetic thickness of $t_1 - t_2$. This effective thickness can be matched to the recording medium without having to use very thin layers t_1 and t_2.

The design rules for a synthetic ferrimagnet sensor free structure are simple. Make the left-hand Co/Fe layer thickness, t_1, be equal to the majority carrier penetration depth, say 75 Å. Make the thickness difference, $(t_1 - t_2)$, appropriate to match the recording medium. The left-hand Co/Fe layer and the pinned layer, or the closest layer of a synthetic ferrimagnet pinned structure, act together to produce the giant magnetoresistive effect. The right-hand Co/Fe layer is merely there to reduce the effective flux capacity of the structure.

Further Reading

Additional reading material is listed here that will prove helpful to readers who seek more detailed information.

Further Reading

Leal, J. L., and Kryder, M. H. (1996), "Interlayer Coupling in Spin Valve Structures," *IEEE Trans. Mag.* **32**(5), 4642–4644.

Veloso, A., and Freitas, P. P. (1999), "Spin Valves with Synthetic Ferrimagnet and Antiferromagnet Free and Pinned Layers," *IEEE Trans. Mag.* **35**(5), 2568–2570.

Meguro, K., Hoshiya, H., Watanabe, K., Namakawa, Y., and Fuyama, M. (1999), "Spin-Valve Films Using Synthetic Ferrimagnets for Pinned Layer," *IEEE Trans. Mag.* **35**(5), 2925–2927.

CHAPTER

13

Antiferromagnet Upsets

When a GMR spin valve suffers a severe thermal asperity, it is possible that the antiferromagnet layer can be heated above its blocking temperature. For example, as is shown in Table 11.1, $Fe_{50}Mn_{50}$ and $Ni_{50}O_{50}$ have blocking temperatures of only 180 and 210°C.

Subsequently, the AFM cools down through its blocking temperature. If the entire AFM is heated over the blocking temperature, the result is that the exchange coupling field, H_{EX}, will have been reversed. The spin valve GMR characteristic will now be unbalanced and the head will show both pulse-amplitude and track-profile asymmetries. Of course, intermediate cases, where only part of the AFM goes above the blocking temperature, can also occur with the result that only part of the pinned layer remains properly magnetized. These phenomena are called antiferromagnet upsets.

Perhaps surprisingly, it is possible to recover from an antiferromagnet upset very quickly in real time with the spin valve remaining installed in the drive. Figure 13.1 shows a simple spin valve with the measuring current, I_M, flowing out of the plane of the diagram. This direction of I_M produces a current bias magnetic field which opposes the AFM–pinned layer exchange field, H_{EX}. Note that in the bias compensator design shown in Fig. 11.10, the measuring current polarity is opposite and the current bias field in this case adds to H_{EX}.

Figure 13.2 shows the waveform of the measuring current that is required to effect recovery from an antiferromagnet upset.

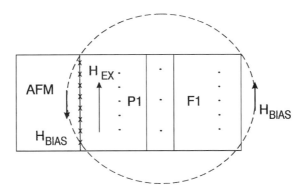

Fig. 13.1. The measuring current vertical bias fields in a simple spin valve.

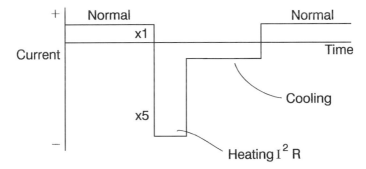

Fig. 13.2. Measuring current waveform for antiferromagnet upset recovery.

Initially, the measuring current, I_M, is at the normal magnitude and polarity for a simple spin valve. It is likely to be causing Joule heating of the spin valve of about 50°C above ambient. Upon detecting that an antiferromagnet upset has occurred, for example by observing that the pulse amplitude asymmetry has suddenly increased, the recovery procedure commences. First, the current polarity is reversed and the amplitude is increased. In Fig. 13.2, I_M is shown changing to $-5\ I_M$. This causes a 25-fold increase in the I^2R heating and, in fact, in thermal equilibrium the temperature rise would now be 25 × 50 = 1250°C above ambient, which is close to the spin valve's melting temperature. This high magnitude current is maintained for a few tens of nanoseconds until the spin valve, and in particular the AFM, is heated to a temperature about 50°C above the AFM's blocking temperature. Next, the current is reduced to $-I_M$. This value is maintained for several tens of microseconds, during which period the AFM cools down through its blocking temperature. In this manner, a proper magnetic annealing procedure has been performed that resets the AFM and restores the AFM-pinned layer exchange field, H_{EX}, Finally, the measuring current is switched back to its normal magnitude and polarity, $+I_M$.

The complete antiferromagnet recovery process takes several tens of microseconds only and during this period, of course, data is not being read properly. Provided the lost data is within the correction capacity of the error detection and correction, EDAC, coding employed, the upset recovery process is transparent to the disk drive controller. If the EDAC capacity is exceeded, then a second reading operation on the next revolution of the disk will, of course, be required.

Further Reading

Additional reading material is listed here that will prove helpful to readers who seek more detailed information.

Williams, Edgar (2001), *Design and Analysis of Magnetoresistive Recording Heads,* John Wiley and Sons, New York, p. 243.

CHAPTER

14

Flux-Guide and Yoke-Type Magneto-Resistive Heads

Two persistent drawbacks encountered with single-element shielded magneto-resistive head designs are their sensitivity to frictional heating effects and the problem of electrical shorting to conductive recording media. When the tape or disk rubs across the top of the head, the frictional force times the rubbing distance is equal to the work performed and heat generated. The thermal coefficient of resistance of permalloy (0.3%/°C) is such that a 10–20°C rise in temperature produces a change in resistance equal to the anisotropic magneto-resistive change ΔR. This phenomenon, termed *thermal spikes* or *asperities*, is well understood. Typically, the heating cycle is very short (a few nanoseconds) and the cooling cycle, controlled by thermal diffusion into the shields, is rather long (a few tens of microseconds). Thermal asperities of lesser magnitude, but of both positive and negative sign, occur much more frequently and are due to changes in the MRH–reading medium spacing and, therefore, thermal conductance. The effect of thermal asperities can be eliminated completely by using some of the double magneto-resistive element designs discussed next in Chapter 15. It can also be minimized by careful design of the spectral content of the channel codes and the postequalization strategies used.

Flux-guide and yoke-type MRHs offer yet another way to eliminate completely thermal asperity effects. This is accomplished by moving the thermally sensitive MRE away from the head–medium contact area at the top of the head.

The problem of the electrical shorting of the MRE measuring current to electrically conductive recording media, such as thin-film rigid disks, has to be addressed by the system designer. It can be ameliorated, for example, by "floating" the disks. Floating means electrically isolating the disks and maintaining them at the same electrical potential as the MREs. The MRE of flux-guide and yoke-type MRHs has to be electrically insulated from the yoke structure in order to prevent the measuring current from spreading into the yoke. This electrical insulation solves the head–disk shorting problem.

Flux-Guide MRHs

A flux-guide design is shown in Fig. 14.1. The MRE spans or bridges a gap in a highly permeable flux guide that is placed between the highly permeable shields S1 and S2.

The mode of operation is self-evident. The MRE senses the flux flowing down the center flux guide just as does the MRE in a conventional design of a shielded MRH. It follows that the output voltage spectral shape and pulse shape of a flux-guide MRH are the same as those of a shielded MRH.

A major disadvantage of this design, and all the other flux-guide and yoke-type MRHs discussed in this chapter, is that the flux efficiency cannot be high. By its very nature, an MRE is a thin film of high magnetic reluctance. Moreover, it is necessary to provide electrically insulating gaps between the MRE and the flux guide.

In Chapter 9, the conventional shielded MRH was discussed as a magnetic transmission line. The flux-guide MRH's flux-guide/shield system

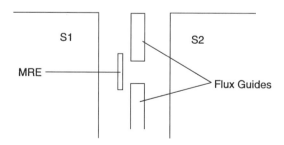

Fig. 14.1. The structure of a flux-guide shielded MRH.

is likewise a flux transmission line. To achieve a high efficiency in a flux-guide MRH, the flux guides should be thick and shallow, so that their reluctances are negligible. On the other hand, it is necessary to keep the shield-to-shield distance $(2g + T)$ small in order to have good spatial resolution (small PW_{50}) so, unfortunately, the flux guide should not, therefore, be thick. Moreover, the depth of the flux guide has to be large enough so that the thermal impulses generated on the top surface have diffused away sufficiently before reaching the MRE. The net result is that it is difficult to achieve flux efficiencies higher than 20%. Accordingly, this flux-guide design produces much lower specific output voltages than does a conventional shielded MRH.

Yoke-Type MRHs

The basic yoke-type MRH design is shown in Fig. 14.2. A deposited thin-film yoke structure is shown with the lower yoke Y1 separated into two parts and bridged by the MRE. Again, by having the MRE some distance from the recording medium contact area and by having the MRE electrically insulated from the yoke, the two problems of thermal spikes and electrical shorting are resolved.

Once more, because the MRE is thin and has electrically insulating gaps, the magnetic reluctance of the structure is high and the flux efficiency of the head is low. Even with careful design, it is difficult to achieve even 20% efficiency in yoke-type MRHs.

The yoke-type MRH senses the ring-head flux directly, and its spectrum is that given in Eqs. (3.3) and (9.5). Where the shielded MRH or inductive head produces the Lorentzian pulse shape $L(x)$ and is proportional

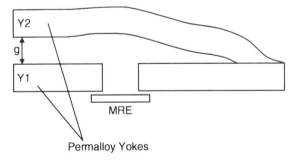

Fig. 14.2. The structure of a yoke-type MRH.

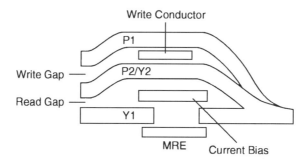

Fig. 14.3. The structure of a thin-film, merged, inductive writing head and yoke-type MRH.

to $dM(x)/dx$, the yoke-type design generates an output which is given by the integral of $L(x)$ and is proportional to $M(x)$. As is discussed in Chapter 19, in some recording systems the inductive head output is integrated and the yoke-type MRH output suits such systems directly.

The complete merged thin-film write head and yoke-type MRH structure used in the Philips DCC is sketched in Fig. 14.3. Again the MRE is the first layer to be deposited, and the order of deposition of the permalloy yokes and poles is Y1, Y2/P2, and P1. Note the placement of the single-turn, servo-bias current conductor and the use of a single-turn write conductor. When fabricating a 16-channel write–read head structure for 4-mm-wide tape, every part of the structure should be made as simple as is possible. A similar philosophy guides the 18- and 36-channel IBM 0.5-in. tape where the merged write heads have two turns only.

Further Reading

Additional reading material is listed here that will prove helpful to readers who seek more detailed information.

Coehoorn, R., Cumpson, S. R., Ruigrok, J. J. M., and Hidding, P. (1999), "The Electrical and Magnetic Response of Yoke-Type Read Heads Based upon a Magnetic Tunnel Junction," *IEEE Trans. Mag.* **35**(5), 2586–2588.

Zieren, V., Somers, G., Ruigrok, J., de Jongh, M., van Straahen, A., and Folkerts, W. (1993), "Design and Fabrication of Thin Film Heads for the Digital Compact Cassette Audio System," *IEEE Trans.* **COMM-29**, 6.

CHAPTER
15

Double-Element Magneto-Resistive Heads

Thermal asperities are the principal disadvantage inherent in the single-element shielded MRHs, be they AMRHs or GMRHs, and they can be alleviated or avoided completely by using structures with two active MREs. Other disadvantages which can be corrected with double-element MRHs are the pulse amplitude asymmetry and side-reading asymmetry inherent to 45° biased AMRHs and electrical shorting to the disk in both kinds of MRHs.

In double-element MRHs, each individual MRE operates as described earlier in this book. The advantages of double-element MRHs are associated with the details of the electrical connections made to, and the directions of magnetization of, the MREs.

Figure 15.1 shows two types of double-element MRHs. In Fig. 15.1A the two MREs are unshielded, whereas in Fig. 15.1B the two MREs are between two shields S1 and S2. It has already been pointed out that the spectral phase of single-element unshielded vertical MRHs and shielded MRHs is the same. Thus, both Eqs. (8.5) and (9.7) contain the factor $j(=\sqrt{-1})$, which denotes a 90° phase shift from the recorded magnetization. It follows that each equation shows the same "digital" output pulse symmetry shown in Figs. 3.4 and 3.5. This pulse basically follows the

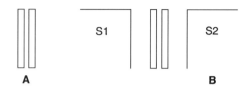

Fig. 15.1. (A) A double-element unshielded MRH and (B) a double-element shielded MRH.

differential of the magnetization transition, $dM(x)/dx$ and has the Lorentzian form,

$$L(x) \propto 1/(1 + (2x/\text{PW}_{50})^2).$$

When two MREs are used, there are only two possible output pulse shapes. When two MRE output pulses of the same polarity are added, the result has the same dM/dx shape. On the other hand, when two MRE pulses of the same polarity are subtracted, the result has a different shape. If we recall the signal processing engineer's method of performing differentiation discussed in Chapter 9, the different shape is bipolar, following d^2M/dx^2 or dL/dx.

Of course, when there are two MREs in proximity to each other, mutual magnetic interaction effects occur and the actual pulses produced are not exactly those given by simple addition and subtraction of the single MRE pulses. The details of the mutual interactions are beyond the scope of this book. Whatever the mutual interactions may be, however, to first order the two possible output pulses of a pair of MREs are dM/dx and d^2M/dx^2.

Classification of Double-Element MRHs

A pair of MREs is shown in Fig. 15.2. Note that in this pair, the magnetization vectors M are parallel and the current flows are parallel. The magnetization vector directions have been established by some vertical bias technique, which is not shown. The current directions are established by means of external conductors, which are also not shown. Finally, the variable parts of the MRE voltage, V_1 and V_2, are presumed to be sensed either in addition, $V_1 + V_2$, or subtraction, $V_1 - V_2$, by suitable electrical connections which, again, are not shown. Subtractive sensing is

Classification of Double-Element MRHs

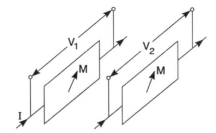

Fig. 15.2. Schematic of two MREs showing their current and magnetization directions and the MR voltages developed.

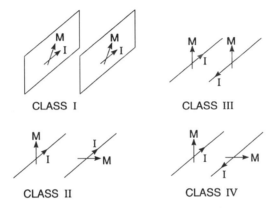

Fig. 15.3. Four classes of double-element MRHs, based on the possible symmetries of the current and magnetization directions.

often called "differential" sensing for obvious reasons. Whereas additive sensing can be accomplished using just two terminals, subtractive or "differential" sensing requires that the head be a three-terminal device.

With two possible current directions, two possible magnetization directions, and two voltage sensing modes, there are eight combinations to consider. The four magnetization and current direction possibilities are shown in Fig. 15.3 and they are classified here, arbitrarily, as Classes I, II, III, and IV. The eight possible combinations are presented in Table 15.1.

Note that the Class I addition configuration yields the usual unipolar output pulse dM/dx or the Lorentzian $L(x)$. Class I subtraction heads have been discussed since the mid-1970s by IBM, and they are called "gradiometer" heads. They produce d^2M/dx^2 or dL/dx bipolar pulses. The Class II addition connection scheme is called, by Kodak, the "dual

TABLE 15.1

Class	Electrical sensing	
	Addition	Subtraction
I	$V = \dfrac{dM}{dx}$	$\dfrac{dV}{dx} = \dfrac{d^2M}{dx^2}$
II	$\dfrac{dV}{dx} = \dfrac{d^2M}{dx^2}$	$V = \dfrac{dM}{dx}$
III	$V = \dfrac{dM}{dx}$	$\dfrac{dV}{dx} = \dfrac{d^2M}{dx^2}$
IV	$\dfrac{dV}{dx} = \dfrac{d^2M}{dx^2}$	$V = \dfrac{dM}{dx}$

magneto-resistor" (DMR) head, and it has the bipolar d^2M/dx^2 type output pulse. The Class II subtraction arrangement gives the normal unipolar dM/dx pulse and is called, by Hewlett-Packard, the "dual stripe" head.

It should be noted particularly that the dual-stripe Class II, subtraction design solves all three of the inherent problems (thermal asperities, pulse amplitude, and side-reading asymmetries) simultaneously. It is a three-terminal device producing the normal output pulse (dM/dx) of double amplitude. It thus seems to be clear that the dual-stripe design is superior to the others.

An example of a possible dual-stripe GMR spin valve structure is shown in Fig. 15.4. It is very similar to the "dual synthetic spin valve" discussed in Chapter 12, but has certain essential or critical differences. The central free layer is now divided into a right and left part by an insulated layer, thus making possible differential or subtractive sensing. Additionally, the outer AFMs and synthetic antiferromagnet pinned layers are now magnetized oppositely, so that the left and right pair of spin valves operate with opposite polarities. The opposite setting of the AFMs can be accomplished by using current bias annealing analogous to that described in Chapter 13 on resetting an AFM layer after an antiferrimagnet upset. In this dual-stripe spin valve, even though there are now 15 layers, it produces two times (200%) the single spin valve output voltage.

It is of interest to note that the remaining four entries in Table 15.1 have not yet received names.

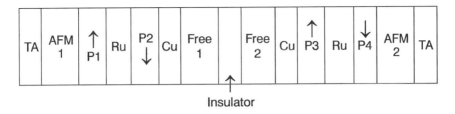

Ta25, NiMn150, CoFe30, Ru8, CoFe25, Cu20, NiFe50,
Al$_2$O$_3$100, NiFe50, Cu20, CoFe25, Ru8, CoFe30, NiMn150, Ta25

15 LAYERS !

Fig. 15.4. A dual-stripe shielded GMR spin-valve MRH.

Advantages of Double-Element MRHs

All of the configurations of double-element MRHs which use subtractive or differential sensing are inherently robust against thermal asperities. This is because the subtractive sensing rejects all common-mode resistance changes such as those caused by the (almost) synchronous thermal fluctuations of the resistance in the MREs.

The configurations which eliminate the even harmonic distortion or pulse amplitude asymmetry are Class II and Class IV. To achieve pulse amplitude symmetry, it is necessary to have differing magnetization directions.

Both Classes II and IV, in which the magnetizations are not parallel, exhibit, to first order, symmetrical side-reading or off-track responses. Should absolutely symmetrical behavior be required, it can be achieved, in principle, by changing the magnetization angles from the pattern shown in Fig. 15.5A to the configuration shown in Fig. 15.5B. Perfect off-track symmetry is expected in all of the four possible patterns which have their z and y magnetization components antiparallel and parallel, respectively, as occurs in Fig 15.5B.

The configurations in which the current in the MREs flows in opposite directions (Classes III and IV) seem attractive. They offer an easy way to alleviate, but not eliminate, the electrical shorting to disk problem. When one end of each MRE is connected and held at zero, or ground, potential, the average potential of the complete double-element structure is also zero. This, of course, means the device requires three external connections or terminals. The same zero average potential result can, of course, be

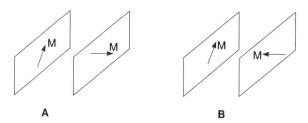

Fig. 15.5. (A) Up–down magnetization giving low side-reading asymmetry and (B) up–down magnetization giving zero side-reading asymmetry.

achieved when the currents flow parallel, but in this case four external connections are needed.

General Comments

Double-element MRHs offer many advantages over single-element MRHs. These advantages are achieved at the expense of considerably increased device and circuit complexity. It follows then that the system designer has to make choices. These choices can only be made after obtaining knowledge of the relative importance of the problems encountered in particular recording systems. For example, electrical shorting and side-reading symmetry might be of little importance in a relatively wide track-width tape recorder using a magnetic coating of low electrical conductivity. Thermal asperities (TAs) might be considered of little importance in flying-head disk drives because of their relative infrequency.

Adding to the difficulty of selecting the best design are, naturally, the complicated issues of device fabrication and process yields, both of which are beyond the scope of this book.

Nevertheless, some general comments are possible. For example, the question of whether the unipolar Lorentzian pulse $L(x)$ is to be preferred over the bipolar Lorentzian $dL(x)/dx$ depends on the origin of the major noise sources in the recording system. If the major noise source is the recording medium itself, then there is little to choose between the two possibilities because either pulse can be transformed, by extremely simple external circuitry, into the other. Moreover, the transformations (differentiation and integration) are linear and reversible so that no information is, in principle, lost in the process.

When, however, the major noise source occurs later in the reading cascade, for example, in the reading head or first-stage read amplifier, the ability to perform signal transformations by means of the spatial design of the reading head before the noise can be extremely useful. An analogous situation arises in the yoke-type (DCC) MRHs discussed in Chapter 14, where the output is proportional to $M(x)$ or the integral of $L(x)$.

Some of the configurations shown in Fig. 15.3 appear to be easier to implement than others. For example, in the Class II and III configurations, the magnetization directions can be established by using MRE current bias alone. Other configurations require the provision of vertical bias by one of the other techniques outlined in Chapter 6.

Finally, it should be noted that all of the configurations that yield the d^2M/dx^2 output can, in principle, be operated satisfactorily without shields. Recall that the fundamental electrical shunting problem of Hunt's unshielded vertical head causes the high-frequency output spectrum to roll off as $1/k$. A second spatial differentiation, which achieves the change from a dM/dx to a d^2M/dx^2 response, is the equivalent of multiplying the spectrum by the factor jk. Obviously, the short wavelength spectrum roll-off is corrected. The remaining factor, j (a 90° phase shift), can then be corrected with an external filter, called an *all-pass phase shifter* or *Hilbert transformer*.

The study of double-element MRHs is a very active area, rife with invention. Time alone will reveal whether or not there is a preferred configuration for specific applications.

Further Reading

Additional reading material is listed here that will prove helpful to readers who seek more detailed information.

Indeck, R. S., Judy, J. M., and Iwasaki, S. (1988), "A Magnetoresistive Gradiometer," *IEEE Trans.* **MAG-24**, 6.

Mallinson, J. C. (1990), "Gradiometer Head Pulse Shapes," *IEEE Trans.* **MAG-26**, 2.

Smith, N., Freeman, J., Koeppe, P., and Carr, T. (1992), "Dual Magnetoresistive Heads for High Density Recording," *IEEE Trans.* **MAG-28**, 5.

CHAPTER

16

Comparison of Shielded Magneto-Resistive Head and Inductive Head Outputs

It is of great interest to compare the output voltages produced by shielded MRHs and inductive heads because the decision to use a shielded MRH rather than an inductive head is frequently dependent on this comparison. There are two ways to make this comparison. In the first, the ratio of the small-signal output voltage spectra of a shielded MRH and an inductive head is considered. This ratio is principally of interest in linear, analog recording.

The second comparison considers the case of a shielded MRH in which the MRE element thickness T and depth D have been chosen so that the peak voltage of the isolated transition output pulse in digital recording is optimized. Examples of this design process are discussed in Chapters 17 and 18.

Small-Signal Voltage Spectral Ratio

The small-signal output voltage spectrum of the single-element shielded GMRH is, following Eq. (9.7),

Small-Signal Voltage Spectral Ratio

$$V_{MRH}(k) = \frac{1}{2}\left(\frac{I\Delta R}{2}\right)\left[\frac{4\pi M_R(1-e^{-k\delta})e^{-kd}}{k}\right]\left(\frac{1}{B_S T}\right)$$

$$\times \left(\frac{\sin kg/2}{kg/2}\right) 2j\sin\left(\frac{k(g+T)}{2}\right). \qquad (16.1)$$

The small-signal spectrum of the shielded AMRH is the same but with $(I\Delta R)$. The output voltage spectrum of an N-turn inductive head operated at a head-to-medium relative velocity V is, from Eq. (3.5),

$$V_{IND}(k) = j10^{-8} NVW[4\pi M_R(1-e^{-k\delta})e^{-kd}]\left(\frac{\sin kg/2}{kg/2}\right). \qquad (16.2)$$

When the common terms are cancelled, the ratio of the shielded GMRH to inductive head output voltage is

$$\frac{V_{MRH}}{V_{IND}} = \frac{\frac{1}{2} \cdot \frac{I\Delta R}{2} \cdot \frac{1}{k} \cdot \frac{1}{B_S T} \cdot \frac{2\sin k(g+T)}{2}}{10^{-8} NVW}, \qquad (16.3)$$

and similarly for the AMRH but with $(I\Delta R)$. This ratio can be evaluated by applying the following approximations.

The measuring current is usually limited by the electromigration phenomenon discussed in Chapter 17 and is $J_0 TD$, where the current density is in the range 10^7–10^8 A/cm^2. The effective magnetoresistive coefficient, J_0, is taken to be 2% in the case of AMR heads and 8% for GMR spin valves. The saturation induction and resistivity of permalloy are 10,000 gauss and 20×10^{-6} Ω-cm, respectively. Finally, for simplicity, it is assumed that $g \approx 4T$ and that the product $k(g+T)$ is small.

After making all the preceding substitutions, Eq. (16.3) becomes

$$\frac{V_{AMRH}}{V_{IND}} \approx \frac{10^{-2} J_0}{NV} \qquad (16.4A)$$

and

$$\frac{V_{GMRH}}{V_{IND}} \approx \frac{4 \times 10^{-2} J_0}{NV}. \qquad (16.4B)$$

These equations can be restated in a particularly direct form:

$$V_{AMRH} \equiv V_{IND}(NV \approx 10^{-2}J_0 \text{ cm/sec}) \qquad (16.5\text{A})$$

and

$$V_{GMRH} \equiv V_{IND}(NV \approx 4 \times 10^{-2} \text{ cm/sec}) \qquad (16.5\text{B})$$

AMRHs and GMRHs have output voltage equal to that of an inductive head with a turn–velocity product of $NV = 10^{-2}J_0$ and $4 \times 10^{-2}J_0$, respectively.

Consider an analog audio recorder such as the Philips Compact Cassette, where the tape speed is about 10 cm/sec. If the current density is $J_0 = 10^7$ A/cm^2, the shielded AMRH small-signal sensitivity is equivalent to that of an $N = 10^4$ turn inductive head. At a tape speed of 100 cm/sec, equivalence occurs with an $N = 10^3$ turn inductive head. Clearly, as Hunt pointed out, AMRHs are extremely sensitive at low tape speeds.

Large-Signal Voltage Comparison

In digital recorders, for example, hard-disk drives, the remanence of the recording medium is usually saturated. It follows that the fields and fluxes emanating from the magnetization transition are not small signals.

In these cases, it is necessary to design the shielded MRH carefully so that the quasi-linear range of operation of the MRE is not exceeded. In terms of magnetization angle changes, typically excursions of ±30° about the quiescent vertically biased 45° are acceptable in AMRHs. In spin valves, dynamic ranges of ±60° are used. Beyond these angles, sufficient amplitude distortion is incurred that it is usually deemed to be unacceptable even for digital recording. As discussed in Chapter 17, the flux capacity of the MRE must match the remanence times thickness product of the recording medium; thus, $B_S T \approx 4\pi M_R \delta$.

Consider a shielded MRH where the MRE's $D \leq l$, the transmission line characteristic length. Of the total magneto-resistive dynamic range only ±50% is available for peak positive- and peak negative-going signals, respectively. If the current density is J_0, the measuring current, I, is $J_0 TD$ and the output voltage $\delta V = I \delta R$. In simple AMRHs δR is about 1% of $R = \rho w/TD$; with GMR spin valves δR is about 4% of R. It should be noted that the depth D and thickness T of the MRE do not influence the output voltage δV in this optimized design case because $R \propto 1/TD$ and $I \propto TD$. The output voltages are

$$_{\text{OPT}}V_{\text{AMRH}} = 2 \times 10^{-7} J_0 W \text{ volts} \quad (16.6\text{A})$$

$$_{\text{OPT}}V_{\text{GMRH}} = 8 \times 10^{-7} J_0 W \text{ volts}. \quad (16.6\text{B})$$

Thus, for example, a soft adjacent layer (SAL) anisotropic MR head operating at 10^7 amps/cm^2 has a specific output (peak) voltage of 2 volt/cm of trackwidth. Many early SAL AMRHs produced much lower specific voltages because an electrically conducting spacer (usually the high resistance (tetragonal) phase of tantalum, $\rho = 180 \times 10^{-6}$ Ω-cm) was employed. Accordingly, the electrical shunting of both the tantalum spacer and the SAL reduced the output voltage, typically to about 30% only. More modern designs use an insulating spacer in order to eliminate the electrical shunting.

As another example, consider a GMR spin valve head using the very high current density of 5×10^7 amps/cm^2. As is discussed further in Chapter 18, such high current densities have become acceptable because not only have the half-gaps, g, become smaller, but also gap materials of significantly higher thermal conductivity are being used. Such a spin valve has a specific output voltage of 40 volts per centimeter of track width. Indeed, this output voltage is frequently stated as 4 kilovolts per meter of track width, thus emphasizing its extraordinarily large magnitude.

When considering shielded MRHs that have been optimally designed, it is not possible to make universal comparisons with inductive heads. Whereas the optimized MRH has a constant specific output voltage, the voltage produced by a linear inductive head is always proportional to the $M_R \delta$ product of the recording medium.

Consider a modern thin-film disk of the kind which achieve areal densities of 10–20 gigabits per square inch. Typically, the disk maximum remanent magnetization, M_R, is 400 emu/cm^3, the coercivity, H_c is 4000 Oe, and the coating thickness, δ, is 125 Å. A simple way of estimating the peak pulse output voltage when using an inductive head is demonstrated in Chapter 17. If the inductive head is of 100% reading efficiency and has N turns and the disk peripheral velocity is V cm/sec, the peak pulse output voltage is, according to Eq. (17.8),

$$_{\text{PEAK}}E_{\text{INDUCTIVE}} = \frac{10^{-8} NV(8\pi M_R \delta W)}{\text{PW}_{50}}. \quad (16.7)$$

If total head–disk spacing is about 1 μ" (250 Å), the slope parameter f is, using Eq. (2.9), approximately 138 Å. The pulse width, PW_{30} is, by Eq. (3.9), approximately 890 Å. Therefore,

$$_{PEAK}E_{INDUCTIVE} \approx 1.4 \times 10^{-5} NVW. \qquad (16.8)$$

If the inductive head has the greatest number of turns likely, $N = 100$, and $V = 1000$ cm/sec, typical of a 10,000 RPM 2.5" diameter disk, the peak output voltage would be 1.4 volts per centimeter of track width. Recalling that the specific output voltage expected of a simple spin valve is approximately 40 volts per centimeter of track width, we see that the GMR spin valve output is, in this particular case, a factor of almost 30 higher.

In this specific case, then, the optimized shielded GMR spin valve head peak pulse output is equal to that of an inductive with an NV product of 3,000,000 cm/sec. So great is the spin valve output that it surpasses the inductive head performance at any realizable head-to-medium velocity. The highest head-medium velocities occur in rotary head tape recorders and are about 6500 cm/sec. At this velocity an inductive head would need to have a totally impractical 460 turns to equal the GMR spin valve. It seems to be safe to say that GMR spin valves produce substantially more peak output voltage than could be realized by an inductive head in any conceivable head–medium interface.

Note carefully that this large signal (output pulse peak voltage) comparison is based upon a particular head–disk interface. The lower the $M_R\delta$, the more attractive become both AMRHs and GMRH spin valves.

A medium with a lower $M_R\delta$ mandates that the optimized design must have both smaller thickness, T, and lower or shallower depth, D. Ideally, the thickness, T, should be directly proportional to $M_R\delta$ and the "stripe" height should vary as $\sqrt{M_R\delta}$. In AMRHs, however, the magneto-resistive coefficient decreases if the MRE becomes thinner, principally because the surface scattering of electrons, which increases ρ_0, becomes proportionately more important. A practical limit in permalloy is about $T \approx 150$ Å where $\Delta\rho/\rho$ has decreased to about 1% only.

In spin valves not only does ρ_0 increase with thinner sensor (free) layers, but also the GMR $\Delta\rho$ decreases when T is less than the majority carrier penetration depth, λ_1. As may be seen in Fig. 5.9, the GMR coefficient becomes less than 2% when a permalloy layer thickness of less than 10 Å is used.

Further Reading

Additional reading material is listed here that will prove helpful to readers who seek more detailed information.

Potter, Robert I. (1974), "Digital Magnetic Recording Theory," *IEEE Trans.* **MAG-10**, 3.

CHAPTER

17

Simplified Design of a Shielded Anisotropic Magneto-Resistive Head

In this chapter, a simplified design procedure for a single-element, anisotropic magneto-resistive effect shielded head is demonstrated by means of a numerical example. This design procedure is included in this book because it shows three things very clearly. First, it shows precisely which physical phenomenon controls the major aspects of a shielded MRH design. Secondly, it demonstrates that in order to obtain optimum performance from a MRH it *must match properly* the magnetic transition in the recording medium. Third, it demonstrates how the chosen units work. Various different units are used in this chapter in accordance with the conventions used in the disk-drive industry in the United States.

The Written Magnetization Transition

The write-head geometry is shown in Fig. 17.1. The recording medium thickness, δ, is taken here to be 2 μin. or 500 Å. The write-to-recording medium flying height, d_w, is assumed to be 2 μin. or 500 Å. The recording medium is assumed to be similar to that of metal-evaporated Co–Ni–P tape and has the following magnetic properties: saturation remanence (M_R) of 200 emu-cm^{-3} ($4\pi M_R \approx 2500$ G) and coercivity (H_c) of 1600 Oe. Such a recording medium is typical of that proposed for 10 gigabit/in.^{2} tape recording.

The Written Magnetization Transition

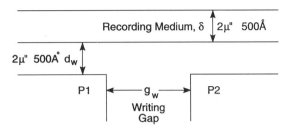

Fig. 17.1. The geometry of the recording medium/write-head interface with the critical dimensions shown.

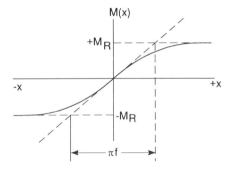

Fig. 17.2. The arctangent magnetization transition.

As was pointed out in Chapter 2, it is not necessary to specify the write-head gap length, g_w, because it is not an important parameter in the writing process. Changes in the write-head gap length are, to a large extent, compensated for by adjusting the write current so that the deep gap field remains close to $3H_c$. In practice, the actual gap length is set by the compromise between improving overwrite (use a larger gap) and decreasing side-writing (use a smaller gap).

In the Williams and Comstock model discussed in Chapter 2, the written magnetization transition is assumed to have the arctangent form shown in Fig. 17.2. The remanent magnetization changes from $-M_R$ to $+M_R$ over a distance of πf, where f is the slope parameter and

$$f = 2\left[\frac{2}{\sqrt{3}}\frac{M_R}{H_c}\delta(d+\delta/2)\right]^{1/2}. \tag{17.1}$$

Substituting $d_w = \delta = 2$ μin., $M_R = 200$ emu-cm^{-3}, and $H_c = 1600$ Oe yields $f = 1.9$ μin.

Inductive and Shielded MRH Output Pulses

An inductive reading head of gap length g_R is shown in Fig. 17.3. The head-to-medium flying height is d_R. This inductive head produces the output pulse shown in Fig. 17.4, which can be characterized by its 50% (of peak) amplitude width, PW_{50}. According to Chapter 3,

$$_{IND}PW_{50} = 2[(d_R + f)(d_R + \delta + f) + g^2/4]^{\frac{1}{2}} \qquad (17.2)$$

In contrast to the write-head gap length, g_R is one of the most critical parameters in system design. A gap length that is too narrow leads to heads of low efficiency. Gap lengths that are too wide produce undesirably wide output pulses. Here, we assume that a 15% increase in PW_{50} is acceptable. Substituting $f = 1.9$ μin. $\delta = d_R = 2$ μin., the pulse widths, $_{IND}PW_{50}$ for gap lengths of 0, 2, 4, and 5 μin. are 9.6, 9.8, 10.3, and 10.8 μin, respectively. Accordingly, an inductive read-head gap length, g_R, of 5 μin. or 0.125 μm is chosen as a satisfactory compromise.

Simple data detection systems, such as the venerable differentiating peak detector, are operated at minimum transition intervals about equal to

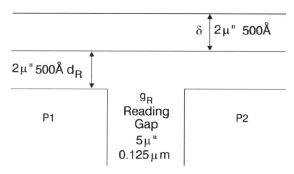

Fig. 17.3. The geometry of the inductive reading head/recording medium interface with the dimensions shown.

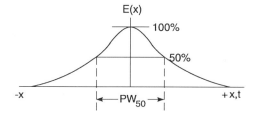

Fig. 17.4. The isolated magnetic transition output voltage pulse shape.

The MRE Thickness T

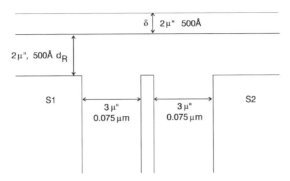

Fig. 17.5. The geometry of the shielded AMRH recording medium interface with the dimensions shown.

PW_{50} yielding, in this case, a maximum flux reversal density of $(PW_{50})^{-1}$ or 90,000 frpi. With a more advanced postequalization and data recovery technique, such as PRML, a 50% increase to 135,000 is feasible. When a channel code such as 2/3 rate (1,7) is used, the bit density is 4/3 the flux reversal density and, accordingly 135,000 frpi corresponds to 180,000 bpi.

In Chapter 9, it was noted that an inductive head with a whole gap length of $1.6g$ has the same spatial resolution as does a shielded MRH with half-gap lengths of g. Thus, a shielded MRH with a half-gap g equal to $5/1.6 \approx 3$ μin. as is shown in Fig. 17.5, produces the same PW_{50} as the 5-μin. gap-length inductive head that was selected earlier.

The half-gap length of 3 μin. or 0.075 μm is the first of the shielded MRH design parameters to be fixed, and it should be noted that it is *intimately* related to the written transition. The MRE thickness T and depth D remain to be determined.

The MRE Thickness T

In order to determine the thickness of the AMR sensor or free layer, we have to ensure that, halfway down the element (at $y = -D/2$), there is sufficient flux from the transition to rotate the magnetization from the quiescent 45° to about 75°. The flux entering the top of the MRE from the transition will be approximately $8\pi M_R \delta W$, regardless of the details of the transition because, as is shown in Fig. 17.6, there are fluxes of magnitude $4\pi M_R \delta W$ flowing into each side of the transition, and, of course, the fluxes flowing into and out of the transition must be equal.

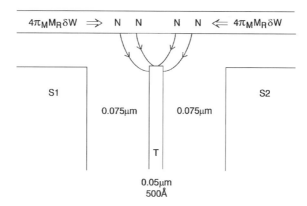

Fig. 17.6. The flux flow into the top of the shielded MRE from a magnetic transition directly over the gap.

As discussed in Chapter 9, the flux falls off of almost linearly with depth, provided the element depth, D, is equal to or less than the characteristic decay length $[Tg\mu/2]^{1/2}$. At the midpoint, then, the flux in the sensor is $4\pi M_R \delta W$. It follows that

$$4\pi M_R \delta W = 4\pi M_S TW (\sin 75° - \sin 45°), \qquad (17.3)$$

where M_S is the saturation magnetization of the MRE. Rearranging, $T = M_R\delta/M_S (\sin 75° - \sin 45°)$ or, numerically, $T = 200.500/800 (0.97 - 0.71)$ and the MRE thickness, T, must be 2 μin. or 500 Å.

When the design includes another highly permeable element, such as a soft adjacent layer (SAL), the MRE must be proportionately thinner. When, as is frequently the case, the flux capacity $B_S T_S$ of the SAL is almost the same as that of the MRE, the minimum MRE thickness in this numerical design becomes 1.0 μin. or 250 Å.

The MRE Depth D

The MRE element depth is determined by the requirement that the depth D be less than l, the magnetic transmission line depth discussed in Chapter 9. If the element is significantly deeper than l, the maximum flux efficiency (≈50%) of the shielded MRH design cannot be realized. If the dimension D is too small, there is the possibility that the design will be needlessly difficult to produce with high process yields.

The transmission line length $l = \{Tg\mu/2\}^{1/2}$, where μ is the low field permeability of the MRE. The permeability of permalloy can be taken to be approximately equal to $B_S/H_k \approx 1600$. With $T = 2$ μin., $g = 3$ μin., $l = 70$ μin. or 1.75 μm. For ease of fabrication, $D = 1.75$ μm is chosen here.

The Measuring Current I

It is, of coursse, desirable to operate the MRH with the highest possible measuring current, I, because the output signal amplitude is directly proportional to I. Two different physical phenomena, however, limit the current magnitude that can be used.

The first operational limit is the I^2R joule heating of the MRE. This heat has to be conducted away by the structure surrounding the MRE. Generally, MREs are operated at temperatures of about 50°C above ambient. If the current were then to be doubled, the temperature rise would then become four times greater (200°C). Joule heating causes the MRE temperature to rise very quickly since most MRHs have thermal diffusion time constants of only a few tens of microseconds.

Nearly all the Joule heating flows into the shields, and it is a simple matter to estimate the temperature rise, θ, of the MRE. The Joule heating is $(J_0TD)^2\rho W/TD$ where ρ is the resistivity of the MRE. The heat flow into the shields is $2\kappa DW\,\theta/g$, where the factor of 2 accounts for the two heat flows (into the shields S_1 and S_2), and κ is the thermal conductivity of the half-gap material. It follows that

$$\theta = J_0^2 Tg\rho/2\kappa. \qquad (17.4)$$

The second limit is set by the phenomenon of electromigration. In electromigration the repeated collisions of the conduction electrons against the scattering centers, which are responsible for the electrical resistivity, eventually causes bulk movement or migration of the conductor's atoms. The kinetic energy of the conduction electrons adds unidirectionally to the randomly oriented thermal energy ($1/2\,kT$ per degree of freedom) which is the root cause of mass diffusion. Electromigration can be considered as electric-field- or current-driven mass diffusion. It becomes faster the higher the electrical current density (J_0 amps/cm^2) and, of course, the higher the temperature. If allowed to occur at too high a rate, the MRE or conductor leads in an MRH will narrow down and eventually open the circuit.

The critical current densities are highest in materials with high melting temperatures such as the so-called "refractory" metals (Pt, Ta, W, etc.), because their atoms are more tightly bound. There is, however, no simple way to calculate the critical current density.

The electromigration phenomenon is insidious because the material transfer depends not only on the current, but also on the accumulated time in service of the MRE. Whereas the Joule heating limits can be established in seconds, the determination of the electromigration limit can take many thousands of hours of device testing. Here it will simply be assumed that $J_0 = 10^7$ amps/cm^2.

When the allowable current is set by electromigration concerns, the measuring current is J_0TD. Substituting $J_0 = 10^7$ A-cm^{-2}, $T = 500$ Å or 5×10^{-6} cms, and $D = 1.75$ µm or 1.75×10^{-4} cms gives $I = 9$ mA.

The Isolated Transition Peak Output Voltage δV

The specific peak output voltage of an *optimally designed* permalloy MRH has already been discussed in Chapter 16. It is about $2 \times 10^{-7} J_0$ V per centimeter of track width W. For convenience, however, the basic arguments leading to this result are repeated here.

Of the total magneto-resistive change (2% in permalloy) only one-half is available for negative and positive pulses. The resistance $R = \rho W/TD$. The measuring current is J_0TD. The peak output voltage is, therefore,

$$\delta V = \frac{1}{2}(10^7 \, TD)\left(\frac{2}{100}\right)(20 \times 10^{-6})\frac{W}{TD} = 2W \text{ volts.} \quad (17.5)$$

Note that when electromigration sets the measuring current density limit, the peak output voltage is independent of the dimensions T and D.

Alternative designs were considered earlier in this chapter, where there are two permeable elements, such as the SAL and the MRE, in the gap. Clearly, the specific peak output voltage is *exactly the same* in these cases, provided the proper changes in MRE thickness and measuring current are made, and an insulating space is used.

Now the simplified design is complete. For a 2500-G remanence, 1600-Oe coercivity medium of 500 Å thickness, the peak pulse output voltage is 2 V per centimeter of track width at a measuring current of 9 mA, when the shielded MRH has half-gaps of 3 µin. and the MRE dimensions are 500 Å thick and 1.75 µm deep. All that remains now is a comparison of the shielded MRH peak output voltage with that of a comparable inductive head.

Optmized MRH versus Inductive Head Voltages

There are, of course, many ways in which the peak pulse output voltage of an inductive head can be determined. When the isolated pulse width PW_{50} is known, however, the method described here is both the simplest and most direct.

Consider the output pulse, shown in Fig. 17.7, with peak voltage E_{peak} and 50% amplitude width PW_{50}. The area, A, under the pulse may be approximated as $1/2 E_{peak} \times 2PW_{50} = E_{peak} \times PW_{50}$. This area is, of course, proportional to the flux change $\Delta\Phi = 8\pi M_R \delta W$ of the digital transition because

$$E(x) = -10^{-8} N\, d\phi/dt = -10^{-8} NV\, d\phi/dx \text{ [volts]} \qquad (17.6)$$

so that

$$A = \int E(x)\,dx = -10^{-8} NV \int \frac{d\phi}{dx} dx = -10^{-8} NV\Delta\phi. \qquad (17.7)$$

It then follows that

$$E_{peak} = -10^{-8} \frac{NV(8\pi M_R \delta W)}{PW_{50}}. \qquad (17.8)$$

Upon substituting $M_R = 200$ emu-cm^{-3}, $\delta = 500\ 10^{-8}$ cms, and $PW_{50} = 10.8$ μin. or 27×10^{-6} cms, the result is $E_{peak} = 9.2 \times 10^{-6}\ NVW$ volts.

It follows that the maximum peak pulse of the optimized MRH (2W volts) is equal to that of an inductive head with an NV product equal to 220,000 cms-sec^{-1}. It should be realized that the actual value of the equivalent NV

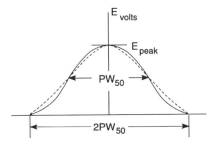

Fig. 17.7. The output voltage pulse of an inductive head and its triangular approximation.

product depends upon the details of the head–medium interface and is not to be regarded as an invariable constant.

Further Reading

Additional reading material is listed here that will prove helpful to readers who seek more detailed information.

Tsang, C., Chen, M., Yogi, T., and Ju, K. (1990), "Gigabit Density Recording Using Dual-Element MR/Inductive Heads on Thin-film Disks," *IEEE Trans.* **MAG-26**, 5.

CHAPTER
18

Simplified Design of a Shielded Giant Magneto-Resistive Head

In this chapter, a simplified design procedure for a single-element, giant magneto-resistive spin valve shielded head is demonstrated by means of a numerical example. As in the previous chapter on shielded AMRHs, this design procedure is included in this book because it shows three things very clearly. First, it shows precisely which physical phenomenon controls the major aspects of a shielded GMRH design. Secondly, it demonstrates that in order to obtain optimum performance from a MRH it *must match properly* the magnetic transition in the recording medium. Third, it demonstrates how the chosen units work. Various different units are used in this chapter in accordance with the conventions used in the disk-drive industry in the United States.

The Written Magnetization Transition

The write-head geometry is shown in Fig. 18.1. The recording medium thickness, δ, is taken here to be 1/2 μin. or 125 Å. The write head-to-recording medium flying height, d_w, is assumed to be 1 μin. or 250 Å. The recording medium is assumed to be similar to Co–Cr–Pt–Ta–B and have the following magnetic properties: saturation remanence M_R of 400 emu-cm^{-3} ($4\pi M_R \approx$ 5000 G) and coercivity H_c of 4000 Oe; such parameters are typical of a disk designed to operate at 10–20 gigabits per square inch areal bit density.

18. Simplified Design of a Shielded Giant Magneto-Resistive Head

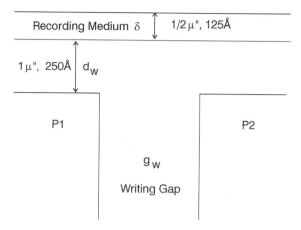

Fig. 18.1. The geometry of the recording medium/write-head interface with the critical dimensions shown.

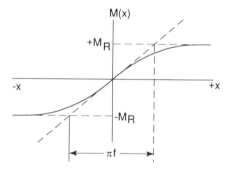

Fig. 18.2. The arctangent magnetization transition.

As was pointed out in Chapter 2, it is not necessary to specify the write-head gap length, g_w, because it is not an important parameter in the writing process. Changes in the write-head gap length are, to a large extent, compensated for by adjusting the write current so that the deep gap field remains close to $3H_c$. In practice, the actual gap length is set by the compromise between improving overwrite (use a larger gap) and decreasing side-writing (use a smaller gap).

In the Williams and Comstock model discussed in Chapter 2, the written magnetization transition is assumed to have the arctangent form shown in Fig. 18.2. The remanent magnetization changes from $-M_R$ to

$+M_R$ over a distance of πf, where f is the slope parameter and

$$f = 2\left[\frac{2}{\sqrt{3}}\frac{M_R}{H_c}\delta\left(d_w + \frac{\delta}{2}\right)\right]^{1/2}. \quad (18.1)$$

Substituting $d_w = 1$ μin., $\delta = 1/2$ μin., $M_R = 400$ emu-cm^{-3}, and $H_c = 4000$ Oe yields $f = 0.55$ μin.

Inductive and Shielded GMRH Output Pulses

A shielded GMR spin valve with half-gaps g is shown in Fig. 18.3. The head-to-medium flying height is d_R. This spin valve head produces the output pulse shown in Fig. 18.4, which can be characterized by its 50% (of peak) amplitude width,

$$_{\text{GMRH}}\text{PW}_{50} = 2[(d_R + f)(d_R + \delta + f) + g^2/2]^{1/2} \quad (18.2)$$

In contrast to the write-head gap length, the half-gap lengths are one of the most critical parameters in shielded MRH design. Half-gap lengths that are too small lead to heads of low efficiency. Half-gap lengths that are too wide produce undesirably wide output pulses. Here we assume that a 15% increase in PW$_{50}$ is acceptable. Substituting $f = 0.55$ μin., $\delta = 0.5$ μin., and $d_R = 1$ μin., the pulse widths, $_{\text{GMR}}$PW$_{50}$, for half-gap lengths of 0, 0.5,

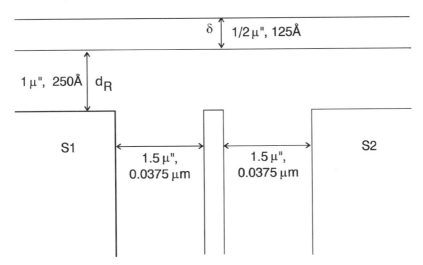

Fig. 18.3. The geometry of the shielded GMRH recording medium interface with the dimensions shown.

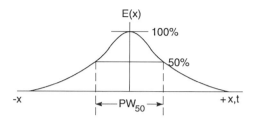

Fig. 18.4. The isolated magnetic transition output voltage pulse shape.

1.0, 1.5 and 2.0 μin. are 3.56, 3.63, 3.83, 4.15, and 4.55 μin., respectively. Accordingly, a half-gap length of 1.5 μin. or 0.0375 μin. is selected as a satisfactory compromise.

Simple data detection systems, such as the venerable differentiating peak detector, are operated at minimum transition intervals about equal to PW_{50} yielding, in this case, a maximum flux reversal density of $(PW_{50})^{-1}$ or 240,000 frpi. With a more advanced postequalization and data recovery technique, such as PRML, a 50% increase to 360,000 frpi is feasible. When a channel code such as 2/3 rate (1, 7) is used, the bit density is 4/3 the flux reversal density and, accordingly, 360,000 frpi corresponds to 480,000 bpi.

The half-gap length of 1.5 μin. or 0.0375 μm shown in Fig. 18.3 is the first of the shielded MRH design parameters to be fixed, and it should be noted that it is *intimately* related to the head–disk and written transition parameters. The MRE thickness T and depth D remain to be determined.

The MRE Thickness T

In order to determine the thickness of the GMR sensor or free layer, we have to ensure that, halfway down the element (at $y = -D/2$), there is sufficient flux from the transition to rotate the magnetization from the quiescent 0° to about 60°. The flux entering the top of the MRE from the transition will be approximately $8\pi M_R \delta W$, regardless of the details of the transition because, as is shown in Fig. 18.5, there are fluxes of magnitude $4\pi M_R \delta W$ flowing into each side of the transition, and, of course, the fluxes flowing into and out of the transition must be equal.

As discussed in Chapter 9, the flux falls off almost linearly with depth, provided the element depth, D, is equal to or less than the characteristic decay length $[Tg\mu/2]^{1/2}$. At the midpoint, then, the flux in the sensor is

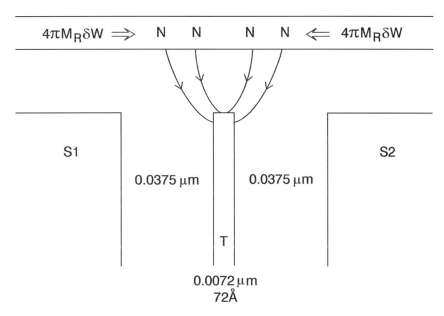

Fig. 18.5. The flux flow into the top of the free layer from a magnetic transition directly over the gap.

$4\pi M_R \delta W$. In order to rotate the free layer magnetization by 60°.

$$4\pi M_R \delta W = 4\pi M_S T W \sin 60°, \qquad (18.3)$$

where M_S is the saturation magnetization of the MRE. It follows that $T = M_R \delta / M_S \sin 60°$ or, numerically, $T = 400.125/800.0.866 \approx 72$ Å. This thickness is also satisfactory from the point of view of maximizing the GMR coefficient because it is comparable to the majority carrier penetration length, λ_1.

The MRE Depth D

The spin valve depth is determined by the requirement that the depth D be less than l, the magnetic transmission line depth discussed in Chapter 9. If the structure is significantly deeper than l, the maximum flux efficiency ($\approx 50\%$) of the shielded MRH design cannot be realized. If the dimension D is too small, there is the possibility that the design will be needlessly difficult to produce with high process yields.

The transmission line length $l = [Tg\mu/2]^{1/2}$, where μ is the low-field permeability of the free or sensor film. The permeability of permalloy can be taken to be approximately equal to $B_s/H_k \approx 1600$. With $T = 74$ Å, $g = 0.0375$ μm, $l = 5000$ Å or 0.5 μm. For ease of fabrication, $D = 0.5$ μm is chosen here.

Now all the principal dimensions, g, T, and D, of the simplified shielded spin valve design are known.

The Measuring Current I

The two phenomena which limit the magnitude of the measuring current, Joule heating and electromigration, have already been discussed in the previous chapter. In spin valves designed for disk files operating at areal densities in the range of 10–20 gigabits per square inch, the current density, J_0 amps per square centimeter, is about five times higher than that used in AMRHs. Typically, then, J_0 is about 5×10^7 amps/cm^2.

Two factors have made this increase possible. First, as has just been shown, the half-gaps used are now considerably thinner and this alone increases the thermal conductance from the spin-valve to the shields. Second, the venerable half-gap material Al_2O_3 has been replaced by AlN. The thermal conductivities of Al_2O_3 and AlN are 1 and 3 watts per meter squared for a temperature gradient of 1°C per meter, respectively. Thinner half-gaps and higher conductivity half-gap material have allowed J_0 to be increased without causing the spin-valve temperature to exceed 50°C above ambient.

The reader is cautioned that there is a world of difference between the J_0 acceptable for the 10- to 20- year lifetime necessary for a satisfactory product and the J_0 that can be used in "laboratory demonstrations" where presumably a lifetime of perhaps 1 week is more than adequate. In some "lab demos" $J_0 > 10^8$ amps/cm^2 have been used!

Assuming that $J_0 = 5 \times 10^7$ amps/cm^2, and that little current flows in the relatively high resistivity AFM, the measuring current in the pinned–spacer–free layer sandwich is approximately 5 mA.

The Isolated Transition Peak Output Voltage δV

The specific peak output voltage of an *optimally designed* spin valve MRH has already been discussed in Chapter 16. For convenience, however, the basic arguments leading to this result are repeated here.

Of the total giant magneto-resistive change (about 8%) only one-half is available for negative and positive pulses. The resistance $R = \rho W/TD$. The measuring current $I = J_0 TD$. The peak output voltage is, therefore, repeating Eq. (17.5),

$$\frac{1}{2}(5.10^7 \, TD)\left(\frac{8}{100}\right)(20.10^{-6})\frac{W}{TD} = 40 \, W \, \text{[volts]}. \quad (18.4)$$

Now the simplified design is complete. For a 5000-G remanence, 4000-Oe coercivity medium of 125 Å thickness, the peak pulse output voltage is 40 V per centimeter of track width at a measuring current of 5 mA, when the shielded MRH has half-gaps of 375 Å and the free or sensor layer dimensions are 72 Å thick and 0.5 μm deep.

Future Projections

It seems to be appropriate, given the astounding rate at which areal densities are increasing (about a factor of two each year) to conclude this chapter with some projections for the future of GMR spin valve heads. In Table 18.1, some predictions published by IBM in 1999 are reproduced.

It is of interest to compare the parameters shown in Table 18.1 with those calculated in the simple design process just discussed. The specific output voltage is only one third that calculated above. Presumably, the GMR coefficient and/or the current density were lower. The half-gaps are larger and the sensor film is both thinner and shallower. Presumably, these differences relate to differing assumptions about the parameters M_R, H_c, and δ of the recording medium.

Several comments are obvious from this table. First, as the areal density increases from 10 to 100 gigabits per square inch, it is anticipated that the linear bit density will increase about fourfold, and consequently both the total (shield-to-shield) read gap and the half-gap lengths must decrease by the same factor of 4. In order to facilitate the fourfold increase in linear density, it is envisaged that the recording medium thickness, δ, and, concomitantly, the free or sensing film thickness must also decrease by a factor of 4.

When both the half-gaps, g, and the sensor film, T, are reduced by a factor of 4, the transmission length l ($\propto \sqrt{Tg}$) and the sensor height, D, must also decrease by a factor of 4 in order to keep the desired 50% read head efficiency. Additionally, a reduction in the flying height by a factor

TABLE 18.1
Future Spin Valve Projections

	10 gbit/in.² head	20 gbit/in.² head	40 gbit/in.² head (projected)	80 gbit/in.² head (projected)	100 gbit/in.² head (projected)
Total read gap	160 nm	113 nm	80 nm	57 nm	40 nm
Half-gap, g	80 nm	57 nm	40 nm	28 nm	20 nm
Read track width, W	0.50 μm	0.35 μm	0.25 μm	0.18 μm	0.13 μm
Sensing film, T	65 Å	46 Å	33 Å	23 Å	16 Å
Sensor height, D	0.40 μm	0.28 μm	0.20 μm	0.14 μm	0.10 μm
Flying height, d	1.0 μ-in	0.71 μ-in	0.50 μ-in	0.35 μ-in	0.25 μ-in
Specific output voltage	1400 μV/μm	1782 μV/μm	2520 μV/μm	3564 μV/μm	5040 μV/μm

Future Projections

of 4 is projected. From Chapters 2 and 3 it can be seen that the result of these changes is to reduce the slope parameter, f, and the isolated pulse width, PW_{50}, also by a factor of 4. When PW_{50} is reduced by a factor of four, then the linear bit density can be increased by a factor of 4.

It is clear that the guiding principle for the projected changes in g, T, D, and d was that of simple scaling. Provided the dynamics (i.e., magnetization rotation) of the GMR spin valve depends solely upon magnetostatics, simple scaling is expected, of course, to be a valid approach. In Table 18.1 even the read track width, W, is scaled by the same factor of 4, giving, in fact, a factor of 16 increase in areal density.

Perhaps of greater interest is the factor of four increase in specific output voltage. It is easy to see why the same factor of 4 occurs. The current in the spin valve is $I = J_0 TD$ and the Joule heating $I^2 R = J_0^2 \rho\, TDW$, where ρ is the electrical resistivity, and this heat must flow into the shields. If the thermal conductivity is κ, the temperature differential between the MRE and the shields is

$$\theta = J_0^2 \rho Tg/2\kappa, \tag{18.5}$$

and it follows that if T and g are both scaled down by a factor of 4, J_0 may be increased by a factor of 4 for the *same* temperature rise, θ. Now, the output voltage $\delta V \approx \rho_{GMR} IR$, where ρ_{GMR} is the GMR coefficient, so that

$$\delta V \approx \frac{\rho_{GMR}(J_0 TD)(\rho W)}{TD}, \tag{18.6}$$

and it is seen that the specific output voltage $\delta V/W$ is directly proportional to J_0. Thus, when J_0 increases by a factor of 4, the specific output voltage increases by a factor of 4 as appears in Table 18.1. The guiding principle was clearly that the spin valve must not be allowed to exceed a particular temperature, θ, above which there is unacceptably rapid electromigration.

It should be noted that other important consequences of these scaling operations are that the peak pulse output voltage (i.e., the specific output voltage times the read trackwidth, W) remains constant and the measuring current, I, is reduced by a factor of 4. Moreover, the sensing film thickness, T, becomes much less than the majority carrier penetration length, λ_1.

Accordingly, in order to realize the projected increase in specific output voltage shown in Table 18.1 it will be necessary to use the three-layer synthetic ferrimagnet type sensor structures discussed in Chapter 12.

Further Reading

Additional reading material is listed here that will prove helpful to readers who seek more detailed information.

Tsang, C., Chen, M. M., Yogi, T., and Ju, K. (1990), "Gigabit Density Recording Using Dual-Element MR/Inductive Heads on Thin-Film Discs," *IEEE Trans.* **MAG-26**, 5.

Gurney et al (1999) "Spin Valve Giant Magneto-Resistive Sensor Materials for Hard Disk Drives," *Datatech Magazine,* December.

CHAPTER
19

Read Amplifiers and Signal-to-Noise Ratios

Magneto-resistive heads can be operated with two types of read amplifier. Although the more common is the high input impedance, voltage-sensing type, it is possible to use a low input impedance current-sensing design. Some of the relative advantages of these two approaches are discussed in this chapter.

Magneto-resistive heads are used because they provide important advantages over inductive reading heads. Two of these advantages, namely, the high output voltage and its independence of MRH-to-recording medium relative velocity, have already been discussed. In this chapter, the principal factors controlling the signal-to-noise ratio (SNR) of a MRH are reviewed.

Magneto-resistive heads are expected to facilitate the attainment of extremely high areal densities ($\approx$100 gigabit/in.^{2}) in the future. This chapter concludes with a short discussion of the specific characteristics of MRHs that encourage these predictions.

MRH Read Amplifiers

The electrical circuit model of an MRH is shown in Fig. 19.1. The fixed part of the resistance of the MRE, R_0, and the resistance, R, and inductance, L, of the connecting leads are shunted by the stray capacitance, C, of the leads. The MRE is represented by a voltage generator, $\delta V = I \delta R$,

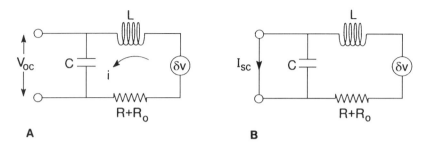

Fig. 19.1. Electrical circuit model of an MRH for (A) voltage sensing and (B) current sensing.

where I is the measuring current and δR is the variable part of the resistance of the MRE. In Fig. 19.1, the measuring current I is not depicted.

When a high internal impedance read amplifier is connected to the head terminals, it measures the open-circuit voltage, V_{oc}, as shown in Fig. 19.1A. The signal current, i, flowing in the loop is

$$i = \frac{\delta V}{(R + R_0) + j\omega L + 1/j\omega C}, \qquad (19.1)$$

and the output voltage is

$$V_{oc} = \frac{i}{j\omega C} = \frac{\delta V}{j\omega(R + R_0)C + (1 - \omega^2 LC)}. \qquad (19.2)$$

The frequency response is second order and has a resonant frequency of $\omega = 1/\sqrt{LC}$. The resonance is undesirable because it causes an oscillating or ringing output voltage. Very similar considerations arise, of course, in voltage sensing with inductive read heads. However, in MRHs the inductance (typically 1 nH) is very much lower than in inductive heads (typically 50 nH), and it follows that the resonant frequency is higher by a factor of approximately $\sqrt{50} \approx 7$. The low inductance of MRHs is due to their "single-turn" structure, and it is another of their many advantages.

A power supply with a high internal impedance is a constant current source. Accordingly, a high-impedance read amplifier is used to produce a constant resistance change measuring current I. This mode of operation is often called the current-forcing, voltage-sensing mode.

The other mode of operation uses a read amplifier that has a low internal impedance, and it is termed the voltage-forcing, current-sensing mode.

The low impedance of the amplifier has the important effect of effectively shunting out the stray capacitance C.

The low impedance amplifier senses the short circuit current, I_{sc}, shown in Fig. 19.1B, where

$$I_{sc} = \frac{\delta V}{(R + R_0) + j\omega L}. \tag{19.3}$$

The frequency response is, therefore, first order, having no undesirable resonance. Instead, the spectrum merely rolls off with a −6 dB frequency $\omega = (R + R_0)/L$. An advantage of this design is the improved high-frequency response.

Current-sensing amplifiers are widely used with inductive heads in digital video recorders, where the operating frequencies (50 to 100 MHz) are higher than in computer peripheral tape recorders. When current sensing is used with inductive heads, the amplifier essentially acts as an integrator and the output follows $M(x)$ rather than the usual $dM(x)/dx$. Integration, which is multiplication by $1/j\omega$ in the frequency domain, occurs because the impedance of the inductive head is mainly inductive, so that $I_{sc} \approx \delta V/j\omega L$. In the case of MRHs, however, the head impedance is principally resistive; $I_{sc} \approx \delta V/(R + R_0)$, and integration does not occur.

Because the MRE is carrying the measuring current I, provision has to be made to control electrical shorting or arcing whenever it touches a conductive recording medium such as a thin-film disk. In some read amplifier designs, the electrical potential of the MRE is held constant, by a servo circuit, so that it is equal to that of the disk. The disk potential can either be ground or some other value. When the disk is "floated," that is, held at a potential other than ground, electrical insulation of the disks, spindle bearings, or motor is, of course, necessary. In "single-ended" read amplifiers, one end of the MRE is simply held at ground potential and disk floating is not needed.

It is obvious from the preceding discussion that there are many different designs for MRH read amplifiers. Which design will be most generally adopted is a question for the future.

Signal-to-MRH Noise Ratios

Throughout this book it has been assumed implicitly that the performance of an MRH is determined by the magneto-resistive coefficient, $\Delta\rho/\rho_0$. In reality, however, this need not always be the case.

Consider the situation where electromigration limits the allowable current density, J_0. In permalloy AMRHs, $J_0 \approx 10^7$ A-cm^{-2} and, in GMR spin valves, $J_0 = 5 \times 10^7$ A-cm^{-2}. The signal voltage, $\delta V = I \delta R$, is simply proportional to $\Delta \rho$, the maximum MR resistance change. If the MRE resistance is R, the mean thermal noise voltage generated is $(4kTR\Delta f)^{1/2}$ and it is proportional to $\sqrt{\rho_0}$. Here, k is Boltzmann's constant, T is the absolute temperature in Kelvins, and Δf is the bandwidth in hertz. It follows that the MRH-limited SNR is proportional to $\Delta \rho / \sqrt{\rho}$ and not the usual $\Delta \rho / \rho_0$.

On the other hand, suppose that in some other MRH design it is joule heating ($I^2 R$) that limits the measuring current, so that I is proportional to $1/\sqrt{R} = 1/\sqrt{\rho_0}$. In this case, the signal voltage $\delta V = I \delta R$ becomes proportional to $\Delta \rho / \sqrt{\rho_0}$ and the MRH-limited SNR is now proportional to $\Delta \rho / \rho_0$.

Consider first the MRH-limited SNR of the optimized shielded AMRH that was designed in Chapter 17. The peak signal voltage is $2W$ volts. The thermal noise voltage is $(4kTR\Delta f)^{1/2}$ where the resistance is $\rho_0 W/TD$ Ω. Substituting $\rho_0 = 20 \times 10^{-6}$ Ω-cm, and the values $T = 250$ Å and $D = 1.75$ µm appropriate to the SAL design, the resistance $R = 4.6 \times 10^4$ WΩ. Assuming Δf, the system bandwidth is 20 MHz, the thermal noise is approximately equal to $1.23 \times 10^{-4} \sqrt{W}$ V rms. The MRH-limited SNR (peak to rms) of the optimized MRH is, therefore, approximately $16{,}250 \sqrt{W}$.

Note that both the MRH noise and the recording medium noise-limited SNR's depend on $\sqrt{W}$. It follows that the ratio of the two SNRs is independent of track width. This is one of the most important reasons for the optimism that MRHs will facilitate extremely high areal density magnetic recording.

Suppose that the optimized shielded MRH of Chapter 17 is used at a track width of $W = 5$ µm. The head-limited SNR is $16{,}250 \, (5 \times 10^{-4})^{1/2} \approx 363$ or 51 dB. This value is much greater than the medium-limited SNR, which may be as low as 20 dB. The conclusion is not only that the system will be recording medium-noise-dominated but also that it will remain so at all other attainable track widths W.

Another way of thinking about the low noise of AMRHs is to ask "At what track width does the MRH-limited SNR fall to the value 30 or 30 dB?" Solving $16{,}250 \, W^{1/2} = 30$ yields $W = 34 \times 10^{-7}$ cm (340 Å). Such a track width is, of course, far below the limits of the photolithographical processes used in making thin-film and MR heads.

It is safe to conclude that an optimally designed shielded AMRH generates so little thermal noise that recording systems can remain medium-noise-limited at all accessible track widths.

An analogous analysis of the MRH-limited SNR of the optimized, shielded GMR spin valve considered in Chapter 18 follows. The peak signal voltage is 40 W volts. The resistance is $0.555 \times 10^6\ W$ ohms. If the system bandwidth is 500 MHz, as is appropriate for a 1 gigabit/second data rate, the Johnson noise is $2.14 \times 10^{-3} \sqrt{W}$ volts rms. The MRH limited SNR is, therefore, $18{,}700\ W^{1/2}$. If this GMR spin valve head is operated at a track width W of 0.5 μm, the peak signal voltage to spin valve rms noise voltage is approximately 132 or 42 dB. Again, since the peak ratio of signal to recording media noise is likely to be only about 20 dB, the conclusion is that this GMR spin valve will easily permit media noise limited operation at 500,000 fipi and 1 gigabit per second.

Magnetization Fluctuation Noise

When the size of the MRE is reduced to the point that thermal fluctuations of the magnetization become of consequence, another form of noise other than the usual Johnson or thermal noise has to be taken into consideration.

The origin of both types of noise can easily be understood by recalling and considering the effect of the equipartition theorem of statistical mechanics. This theorem states that, in thermal equilibrium, every degree of freedom of a system has, most probably, thermal energy equal to $1/2\ kT$. A degree of freedom is understood to mean any part of the system that is able to store energy.

The Johnson noise formula, used earlier in this chapter, can be derived directly from the equipartition theorem. The equivalent circuit of a resistor has inevitably both capacitance, C, and inductance, L, as is shown in Fig. 19.1. In thermal equilibrium the capacitor will store energy, most probably, $1/2\ CV_N^2 = 1/2\ kT$ and the inductor will store $1/2\ LI_N^2 = 1/2\ kT$, where V_N and I_N are the rms noise voltage and current, respectively. In the capacitor this energy is stored over an electrical bandwidth which about equal the reciprocal of its time constant, RC. It follows that if the noise voltage is measured in a narrow band slot, Δf Hz, the noise voltage squared on the capacitor is $kTR\ \Delta f$. In the inductor, the bandwidth is the reciprocal of its time constant, L/R, and the noise current squared is $kT\Delta f/R$. By Ohm's

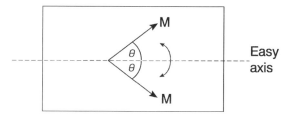

Fig. 19.2. GMR sensor film showing the in-plane magnetization fluctuations.

law this is equivalent to a noise voltage squared equal to $kTR\,\Delta f$ across the resistor. Since we are considering a single resistor these two noise voltages are coherent in time and thus they add as voltages rather than as powers. It follows that the Johnson noise voltage is $2[kTR\Delta f]^{1/2}$ or, as it is usually written, $[4kTR\Delta f]^{1/2}$.

In a similar, simple manner the magnetization fluctuation noise can be estimated directly. The magnetization in the free layer of a spin valve has two degrees of freedom, rotation out of the plane and in the plane. Fluctuations out of the plane of the sensor are small because of the high demagnetization factor (4π), have little effect upon the GMR, and may be ignored. Fluctuation in the plane, shown in Fig. 19.2, on the other hand, causes changes in the giant magnetoresistance proportional to $\sin \theta$, as given in Chapter 10, Eq. (10.1). It is easy to calculate $\sin \theta$ by using the equipartion theorem. The change in magneto-static energy in the sensor layer when the magnetization rotates an angle θ away from the cross-track easy axis is

$$\Delta E = 1/2 M_s H_A V \sin^2\theta \qquad (19.4)$$

where M_s is the saturation magnetization, H_A is the effective anisotropy field, and V is the volume of the layer, WTD.

The effective anisotropy field is

$$H_A = 2K/M_s + M_s(N_y - N_z) + H_{\text{HARD}}$$

where $2K/M_s$ is the induced anisotropy, $M_s(N_y - N_z)$ is the shape anisotropy, and H_{HARD} is the horizontal or hard bias stabilizing field. Here it will be assumed, for the sake of simplicity, that H_A equals 100 Oe.

By the equipartition theorem

$$1/2 M_s H_A V \sin^2\theta = 1/2\, kT \tag{19.6}$$

in thermal equilibrium. Accordingly,

$$\sin\theta = \left[\frac{kT}{M_s H_A V}\right]^{1/2} \tag{19.7}$$

and the most probable or root mean square (rms) GMR noise voltage is

$$\delta V_N = \frac{I\Delta R}{2}\left[\frac{kT}{M_s H_A V}\right]^{1/2}. \tag{19.8}$$

It will be recalled that the normal output pulse peak voltage is also proportional to $I\Delta R/2$, and it follows, therefore, that the peak signal voltage to rms magnetization fluctuation noise ratio is independent of both the measuring current I and the GMR coefficient.

From Eq. (19.7), the most probable magnetization fluctuation angle can be calculated. For the 10–20 gigabit/in.^{2} permalloy sensor considered in Chapter 18, we have $M_s = 800$ emu/cm^3, $T = 72$ Å, $D = 0.5$ µm, and $W = 0.5$ µm the volume is 18×10^{-16} cm^3 and with $H_A = 100$ Oe, the most probable angle is $\sin^{-1} 0.0169$ or $\theta = 0.97°$. Assuming that the noise is flat over a system bandwidth equal to that of the thermal fluctuations, about 10^9 Hz, the peak pulse voltage, due to a magnetization rotation of 60°, to rms magnetization fluctuation noise voltage ratio is $\sin 60/\sin 0.017 = 51.2$ or 34.2 dB. Obviously, for this volume of sensor, the effect of thermal fluctuations is negligible.

When, however, the sensor dimensions are all scaled down by a factor of 4, as occurs in Table 18.1 in going from a 10 to a 100 gigabit/in.^{2} design, the volume decreases 64-fold, the most probable fluctuation angle increases slightly more than 8-fold to 7.79° and the 1 GHz bandwidth SNR decreases to 6.38 or 16.1 dB. In this case, the magnetization fluctuation noise is likely to be dominant.

It is of great interest to compare the magnetization fluctuation noise with the Johnson noise, and in this case considering the relative powers is appropriate, because the two kinds of noise are incoherent. In Table 19.1, these noises are given for two cases, the 10–20 gigabit/in.^{2} spin valve designed in Chapter 18 and a potential 100 gigabit/in.^{2} design with a factor of four scaling applied as in Table 18.1. Note that in this table the

TABLE 19.1
Johnson and Magnetization Fluctuation Noise Powers of 10–20 and 100 gigabit/in.^{2} Spin Valves

	10–20 gigabit/in.^{2}	100 gigabit/in.^{2}
Read track width, W	0.5 μm	0.125 μm
Sensor film, T	72	18
Sensor height, D	0.5 μm	0.125 μm
sin θ	0.0169	0.135
Anisotropy energy barrier, ΔE	3478 kT	54 kT
Resistance, R	28 Ω	112 Ω
Measuring current, I	1.8 mA	0.45 mA
Bandwidth	500 MHz	500 MHz
Johnson noise power	2.32×10^{-10} watts	9.28×10^{-10} watts
Magnetization fluctuation noise power	5.75×10^{-10} watts	368×10^{-10} watts
Total head noise power	8.1×10^{-10} watts	377×10^{-10} watts

system bandwidth is taken to be one half of the magnetization fluctuation bandwidth, so that the fluctuation noise power given is only $1/2(\delta V_N)^2$.

Many fascinating factors are revealed in Table 19.1. First, it is seen that even with the larger-volume (18 × 10^{-16} cm^3) 10–20 gigabit/in.^{2} sensor, the magnetization fluctuation noise power is about twice the Johnson noise. With the smaller (0.28 × 10^{-16} cm^3) 100 gigabit/in.^{2} proposed sensor, the magnetization fluctuation noise is much larger (by a factor of four to the *third* power) and becomes almost a factor of 40 larger than the Johnson noise. This fact is perhaps not surprising since the total anisotropy barrier in the smaller sensor is only 54 kT, which is strictly comparable to the energy barrier, ΔE, of the near-superparamagnetic single-domain grains found in modern thin-film disks. Finally, we may calculate the ratios of peak pulse output power to spin valve total noise power for the two heads. They are 604 (27.8 dB) and 11.4 (10.6 dB), respectively. Clearly, from the noise and SNR viewpoint, the 10–20 gigabit/in.^{2} spin valve is acceptable without reservations because it will permit the system SNR to be medium noise dominated. On the other hand, the 100 gigabit/in.^{2} proposed design should be regarded as only marginally acceptable because the system SNR will now likely be dominated by spin-valve read noise.

Conclusions

Magneto-resistive heads, after early and unsuccessful attempts in analog recording, are now used in all of the many types of digital recorders. There is widespread optimism that anisotropic MRHs will make possible recording at 10 gigabit/in.^{2} in tape drives and that giant MR spin valve heads will allow areal densities approaching 100 gigabit/in.^{2} in disk drives.

To understand why this revolution in reading heads has occurred, it is this writer's opinion that the reader can do no better than to review, once again, the several advantages of MRHs, as propounded by Hunt in 1970. Indeed, the first paragraph of Hunt's paper, given here in Chapter 8, stands as a model of clarity, brevity, and scientific accuracy.

Further Reading

Additional reading material is listed here that will prove helpful to readers who seek more detailed information.

Grochowski, E., and Thompson, D. (1994), "Outlook for Maintaining Areal Growth Rate in Magnetic Recording," *IEEE Trans.* **MAG-30**, 6.

Klaassen, K. B., and van Pedden, J. C. L. (1995), "Read/Write Amplifier Design Considerations for MR Heads," *IEEE Trans.* **MAG-31**, 2.

Mallinson, J. C. (1969), "Maximum Signal-to-Noise Ratio of a Tape Recorder," *IEEE Trans.* **MAG-5**, 3 (also in R. M. White; see Bibliography).

Mallinson, J. C. (1991), "A New Theory of Recording Media Noise," *IEEE Trans.* **MAG-27**, 4.

Murdock, E. S., Simmons, R. F., and Davidson, R. (1992), "Roadmap to 10 Gbt/in.^{2} Media," *IEEE Trans.* **MAG-28**, 5.

CHAPTER
20

Colossal Magneto-Resistance and Electron Spin Tunneling Heads

Despite the overwhelming success and adoption of giant MR spin valve heads and the optimism that they will suffice up to areal densities close to 100 gigabit/in.^{2}, research efforts continue to find even better reading heads. The current activity falls into two areas: so-called colossal magneto-resistance (CMR) and electron spin magnetic tunneling junction (MTJ) heads. The current (2001) status of these two topics is covered in this chapter. It should be clearly understood that, as is the case in all ongoing research activities, no definite conclusions can presently be made about these technologies. Any future outlooks given here represent merely this writer's current opinions and they may well be proven subsequently to be incorrect.

Colossal Magneto-Resistance

Colossal magneto-resistance was first discovered in 1994 at Bell Labs. It was found that thin films of a manganate, $La_{0.67}Ca_{0.33}MnO_x$, displayed a negative, isotropic magneto-resistance effect more than *three* orders greater than the GMR in superlattice films. The films were grown, epitaxially, about 1000–2000Å thick, by pulsed laser ablation on $LaAlO_3$ substrates and were then heat treated at temperatures in the range 500–700°C. The CMR effect causes resistance changes of 127,000% at liquid nitrogen temperatures (77 K) and, more importantly, about 1300% at room temperature. The effect was, with considerable justification, termed the colossal MR effect. In this first work, an extremely large magnetic field, 60,000 Oe (6 tesla), was needed to cause the full resistance change and, just as was the case with the discovery of GMR in 1987, there seemed to be little hope of using CMR in magnetic recording heads.

Intensive work at laboratories throughout the world has now, however, managed to get the fields required down to a few thousand oersteds. This has been achieved by using manganate compositions with lower anisotropy fields ($2K/M$) and by preparing samples with fewer structural defects. Samples are made now by laser ablation, sputtering, and even "floating zone" single-crystal growth processes.

There still exists considerable uncertainty about the exact origin of the CMR. It is known, however, that the manganates have a perovskite-like layered structure. The magnetization is known to be ferromagnetically ordered in the plane of the layers and, depending upon the magnitude of the external field and the temperature, to be either ferromagnetically or paramagnetically ordered from layer to layer. The CMR effect is associated with the magnetic phase change from ferro to para in the layer to layer coupling. The behavior of these oxide films is often described as being "half-metallic," which refers to the fact that half say the spin up, but not the spin down, outer electrons of the manganese atoms are itinerant, mobile, and electrically conducting. Despite all this knowledge, however, a precise theory of CMR appears to remain elusive.

From the device standpoint, CMR in manganates appears to be accompanied by several severe drawbacks. First, in order to be useful in reading heads the magnetic field required remains, at several thousands of oersteds, much too high. In order to equal the sensitivity ($\Delta H\ dR/dH$) of GMR spin valves, the required fields must be reduced to lower than 100 Oe.

Another troubling implication of the high magnetic field needed to cause the ferro- to paramagnetization phase transition is that the permeability is very low ($\mu < 10$). It will, accordingly, be difficult to incorporate these materials in shielded head designs because the transmission line length ($[Tg\mu/2])^{1/2}$ will be very small.

Finally, as oxides, albeit "half-metallic," the electrical resistivities are troublingly high at about 100 Ω-cm. This means that the measuring currents will have to be much lower than in current permalloy GMR heads. When the resistivity is $10^2/20 \times 10^{-6} = 5 \times 10^6$ higher, it follows not only that the

Johnson noise will be excessive but also that measuring currents as low as a few tens of nanoamps are likely to be used. At a data rate of 1 gigabit/second, such a small current is only a few tens of electrons per bit cell and, consequently, the shot, or electron counting, noise problems in the bit detector will likely be unacceptable.

In summary, despite the considerable progress made in the past seven years, CMR based reading heads still appear to be an unlikely possibility, unless the magnetic field required and the electrical resistivity can both be reduced by several orders of magnitude.

Magnetic Tunneling Junctions

The quantum, or wave, mechanical tunneling effect has been known and well understood since the 1930s. It is based on the fact that there is a nonzero probability that an electron can exist in a vacuum outside a conductor. In quantum mechanical terms, this probability is equal to the magnitude squared of the electron's wave function. In wave mechanical terms, the phenomenon may be considered as an "evanescent" wave whose amplitude decays rapidly (exponentially) with distance from the surface.

In the mid-1990s, it was discovered that the magnitude of the tunneling current, in the gap between two ferromagnetic metals, is dependent upon the electron's spin directions or polarizations.

In a prototypical, electron spin polarization modulated, magnetic tunneling junction, the gap between the two ferromagnetic layers is in the range 5 to 10 Å only and is, most usually, Al_2O_3. The tunneling current flows perpendicular to the plane of the ferromagnets, as is shown in Fig. 20.1.

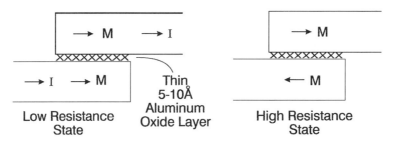

Fig. 20.1. The prototypical MTJ showing the out-of-plane tunneling current direction.

One of the ferromagnets is exchange coupled to an antiferromagnet and is, as in GMR spin valves, called the pinned or reference layer. The other ferromagnetic layer is the free, or sensor, layer and, of course, it will require longitudinal, or hard, bias to hold it single domain. Overall, it will be realized that an MTJ is similar to a GMR spin valve but with an insulating spacer. Accordingly, it is easy to see how an MTJ can be built between shields S1 and S2 just like a spin valve.

The tunneling magneto-resistance ratio (TMR) is typically in the range of 20–40% at room temperature. It will be realized that this is significantly (about 4–5 times) higher than in current GMR spin valves, and this is, of course, one of the principal motivations for investigating MTJs.

Naturally, MTJs come with a wide variety of problems. For many years, the electrical resistance of the Al_2O_3 spacer layers was much too high to allow the design of a low Johnson noise 100 gigabit/in.^{2} reading head. It has become customary to quote the resistance on the basis of perpendicular current flow through a 1 μm square, thus taking the thickness of the layer (5–10 Å) into account implicitly. For many years, the Al_2O_3 layer was deposited by sputtering and the areal resistivities, which depend, of course, exponentially upon the thickness, were typically 10^2, 10^3, 10^4, and 10^5 Ω-μm^2 for thicknesses of 4, 6, 8, and 10 Å, respectively. Consider a 100 gigabit/in.^{2} MTJ of planar dimensions 0.125 μm square and 8 Å in thickness. The resistance is 64 10^4 Ω. In comparison with the GMR spin valve head considered in Chapter 19, which had a resistance of about 100 Ω, this MTJ head would produce 80 times the Johnson noise voltage and, with but 4–5 times the signal voltage, it would be of little practical utility. This fact, of course, leads designers to consider thinner films, but it is very difficult to produce 4 Å thin films without pinholes.

In the last year, however, this unpromising scenario has changed totally. It has been discovered that by using "native" oxide films, the areal resistances found can be much lower, in the range 20–30 Ω μm^2, with even 5 Ω-μm^2 having been reported in early 2001. A native Al_2O_3 film means that instead of sputtering Al_2O_3 directly, first a layer of Al metal is sputtered and this is then oxidized in an oxygen plasma and annealed, typically at 250°C for 1 hour. The result is not only a drastic reduction in areal resistivity but also far fewer pinholes. Apparently, as one might have guessed, it is easier to produce structural defect free and pinhole free films in a layer of low melting temperature metal, Al, than is the case with a refractory oxide such as Al_2O_3.

When the area resistivity is only 20 Ω-μm^2 (at 7 Å thickness), the 100 gigabit/in.^{2} MTJ sensor resistance is only 64 × 20 equals 1280 ohms. Such an MTJ sensor would have about a factor of 10 (10 dB) more Johnson noise power than that of the GMR spin valve design. If pinhole-free, native Al_2O_3 films can be produced at less than 5 Å thickness, the MTJ noise power is expected to be lower than that of the GMR spin valve. MTJs as small as 500 Å square, made by ion beam milling, have been investigated. A typical MTJ structure is IrMn/CoFe/AlO$_x$/NiFeCo.

The technology is presently rife with invention. Amongst the ideas currently being pursued are "dusting" the ferromagnet–insulator interfaces with, for example, Co and Ni; AlN insulating spacers; and the use of CMR maganate/insulator/manganate structures and even semiconductor/insulator/semiconductor structures.

Overall, the future outlook is that MTJ reading heads with both higher signals and higher signal-to-noise ratios suitable for 100 gigabit/in.^{2} areal densities can be made. Whether the performance of such MTJs will prove to be sufficiently stable when operated at, say 100°C (about 50° above ambient in a disk drive) over periods as long as 10–20 years is a matter yet to be determined. Ensuring that there is virtually no interdiffusion, into the ferromagnets, of an insulator layer only two or three atoms (4–5 Å) in thickness is, indeed, a matter of nothing less than atomic engineering.

Further Reading

Additional reading material is listed here that will prove helpful to readers who seek more detailed information.

Junichi, F., Ishi, T., Mori, S., Matsuda, K., and Ohasi, K. (2001), "Low Resistance Magnetic Tunnel Junctions with Various Interface Structures by UHV Sputtering," paper HE-02, Intermag Conference.

Mahesh Samant and Stuart S. Parkin (2001), "Influence of Co and Ni Nano-Interface Layers on Magnetic Tunnel Junction Properties," paper HE-03, Intermag Conference.

Butler, W. H., Zhang, X.-G., Schulthess, T.C., and MacLaren, J. M. (2001), "Theory of Spin-Dependent Tunneling," paper HE-06, Intermag Conference.

CHAPTER
21

Electrostatic Discharge Phenomena

Magnetorestive heads are vulnerable to damage caused by the Joule heating occurring during electrostatic discharges (ESDs). The damage incurred can range from a mere antiferromagnet upset to complete vaporization (boiling) of the MRH. The materials used in MRHs are themselves considered to be resistant to ESD damage because they have high melting and boiling temperatures and also because they have relatively high specific heats, calories per cm^3. However, the thermal capacity, that is, the product of specific heat and volume, is extremely low, particularly in spin valves designed for operation at very high areal densities, because the dimensions, T, D, and W, of the device are so small, as is displayed in Table 18.1.

Electrostatic discharges are familiar to all who live in climates with very low relative humidities. The "zap" which unfortunately accompanies a handshake is an ESD event. The American Society of Testing Methods (ASTM) has developed the equivalent electrical circuit of a human being, shown in Fig. 21.1. It consists simply of a 50 pF capacitor in series with a 1000-Ω resistor.

The uncomfortable "handshake zap" corresponds to a charging voltage on the capacitor of about 20,000 volts. Since the dielectric breakdown electric field in air is about 30,000 volts per centimeter, the spark in the "zap" can occur at spacings greater than half a centimeter.

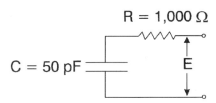

Fig. 21.1. The ASTM "human" model.

In the case of AMR heads, the charging voltage in the human model that causes complete vaporization of the MR structure is typically only 100 volts. Partial damage can be caused with only 10 volts. In many ways, complete destruction is preferable to partial damage because MRHs that have "open circuited" are easy to detect. Partially damaged heads must be measured in order to determine if they should be used.

It is the energy, $\frac{1}{2}CV^2$, stored in the "human" model capacitor that provides the energy required to melt or vaporize the MRH. If an MRH of resistance R ohms is connected to the "human" model, the fraction of the capacitor's stored energy that appears in the MRH is, of course, $R/(R + 1000)$. If the specific heat, the melting (or boiling) temperature, and the volume of the MRH structure is known, the damage threshold voltage can be estimated easily.

In the case of GMR spin valves, the thermal capacity is usually much lower than in AMRHs. The complete vaporization voltage can be as low as 10 volts in the 10–20 gigabit/in.^{2} spin valves used today. Even a capacitor voltage of 1 volt can cause partial damage.

With such spin valves, great care is required in the selection of any electronic volt (or ohm) meters that are used to test the heads for ESD damage. In many commercial ohmmeters, the very act of making contact with the probes causes the discharge of a (small) capacitor within the measuring instrument itself. The result, in the worst case, can be that every head being tested is destroyed. It behooves the user to study the operator's manual and circuit diagram of the instrument carefully or consult a competent electrical engineer.

The measures that can be taken to minimize the likelihood of ESD events are obvious, when it is realized that the very existence of the electrostatic charge depends upon insulators, just as the handshake "zap" depends upon a low humidity and, thus, insulating atmosphere. Accordingly, ESD can be avoided by making everything conductive and be electrically bonded together. The precautionary steps to be taken are widely used, of course, in other industries, for example, in semiconductor device production, and include the following:

> Use ionized air clean benches
> Keep the relative humidity high
> Use conductive (polynitrile) gloves
> Ground wrists and ankles to the floor
> Use conductive clothing
> Use conductive floor coverings
> Electrically bond all equipment together
> Ship heads in metalized (Au, Al) plastic wrappers
> Ship heads with shorting clips on their terminals

Appendix

Cgs-emu and MKS-SI (Rationalized) Units

This appendix consists of two tables. Table A.1 is a conversion table provided for the immediate conversion of the cgs-emu units used in this book into the corresponding MKS-SI units.

TABLE A.1
Cgs-emu to MKS-SI Units Conversion Table

Quantity	Symbol	Cgs-emu	Conversion factor, C	MKS-SI
Magnetic flux density	B	gauss	10^{-4}	tesla
Magnetic flux	ϕ	maxwell	10^{-8}	weber
Magnetomotive force	mmf	gilbert	$10/4\pi$	ampere
Magnetic field	H	oersted	$10^3/4\pi$	A/m
Magnetization	M	emu/cm^3	10^3	A/m
Magnetization	$4\pi M$	gauss	$10^3/4\pi$	A/m
Specific magnetization	σ	emu/g	1	$A \cdot m^2/kg$
Magnetic moment	μ	emu	10^{-3}	$A \cdot m^2$
Susceptibility	χ	Dimensionless	4π	Dimensionless
Permeability	μ	Dimensionless	$4\pi/10^{-7}$	Dimensionless
Demagnetization factor	N	Dimensionless	$1/4\pi$	Dimensionless

TABLE A.2
See Where the 4π Factor Goes!

	MKS-SI (rationalized)	cgs-emu
Field of a wire	$H = \dfrac{2I}{(4\pi)r}$, A/m	$H = \dfrac{0.2I}{r}$, oersted
Field of a solenoid	$H = \dfrac{NI}{l}$, A/m	$H = \dfrac{0.1(4\pi)NI}{l}$, oersted
Mmf	mmf $= NI$, A	mmf $= 0.1(4\pi)NI$, gilbert
Magnetic moment	$\mu = \int H\,dv$, Am^2	$\mu = \dfrac{1}{4\pi}\int H\,dv$, emu
Magnetization	$M = \dfrac{1}{VOL}\sum \mu$, A/m	$M = \dfrac{1}{VOL}\sum \mu$, emu/cm^3
Flux density	$B = \mu_0(H + M)$, tesla	$B = H + (4\pi)M$, gauss
Permeability of free space	$\mu_0 = (4\pi)10^{-7}$, number	$\mu_0 = B/H = 1$, number
Flux	$\phi = \int B\,dA$, weber	$\phi = \int B\,dA$, maxwell
Voltage	$E = -N\dfrac{d\phi}{dt}$, V	$E = -10^{-8}N\dfrac{d\phi}{dt}$, V

The second table will be of interest to those readers with greater curiosity. Table A.2 gives the defining equations of all the magnetic entities used in this book, for both the cgs-emu and the MKS-SI (rationalized) systems of units. Notice that the difference between the two systems is not only a mere substitution of meters for centimeters and kilograms for grams, but also it includes a purely arbitrary shuffling of the position of the inevitable factor 4π. Accordingly, the second table has been given the title "See Where the 4π Factor Goes!" In both sets of equations, current and voltage are amperes and volts, respectively.

Further Reading

Additional reading material is listed here that will prove helpful to readers who seek more detailed information.

Brown, W. F. (1984), "Tutorial Paper on Dimensions and Units," *IEEE Trans.* **MAG-20**, 1.

Recommended Bibliography on Magnetic Recording

Bertram, H. Neal, *The Theory of Magnetic Recording,* Cambridge University Press, Cambridge, 1994. A highly mathematical and difficult-to-read treatment of the writing, reading, and noise processes in magnetic recording.

Mallinson, John C., *The Foundations of Magnetic Recording,* Academic Press, San Diego, 1993. A nonmathematical, but scientifically accurate, overview of most of the fundamental ideas used in magnetic recording media, heads, and systems.

Mee, C. Denis, and Daniel, Eric D., eds., *Magnetic Recording Handbook,* McGraw-Hill, New York, 1995. A comprehensive collection of contributions, by some 20 experts, on most of the important technologies used in and applications of magnetic recording systems.

Ruigrok, Jaap J. M., *Short Wavelength Magnetic Recording,* Elsevier Advanced Technology, Netherlands, 1990. A highly detailed and easy-to-read survey of both the theory and the practice of high-density magnetic recording.

White, R. M., ed., *Introduction to Magnetic Recording,* IEEE Press, New York, 1985. A most useful collection of reprints of original research papers.

Williams, Edgar M., *Design and Analysis of Magnetoresistive Recording Heads,* John Wiley and Sons, New York, 2001. An incredibly detailed survey of the design of AMR and GMR heads in which most of the abundant data is fitted to closed-form simple algebraic expressions. This book will be found indispensable by those who are actually engaged in designing MR heads.

Index

Additive sensing, 126–127
AlFeSil, 38, 91
All-pass phase shifter, 131
American Society of Testing Methods (ASTM), 171
Ampex, 62
Anisotropic magneto-resistive (AMR) effect
 basis effect, 27–28
 coefficient, 34–35
 permalloy, properties of, 35–38
 sensors or elements (MRE), 28–34
Anisotropic magneto-resistive heads (AMRHs)
 advantages of, 72–74
 Hunt's unshielded horizontal, 74–76
 Hunt's unshielded vertical, 76–78
 invention of, 27
Anisotropic magneto-resistive heads (AMRHs), design of single-element shielded
 inductive and shielded output pulses, 140–141
 isolated transition peak output voltage, 144
 measuring current, 143–144
 MRE depth, 142
 MRE thickness, 141–143
 optimized MRH versus inductive head voltages, 145–146
 written magnetization transition, 138–139
Antidiffusion, 105–106
Antiferromagnets, 5–6
 free layer and synthetic, 115–116
 pinned layer and synthetic, 111–115
 upsets, 118–120
Areal densities, 153

Barber-pole scheme, 58–59, 79
Barkhausen noise, 34, 64
Bell Labs, 166
B field (magnetic flux density). *See* Magnetic flux density (B field)
Bias compensator layer, 108
Blocking temperatures, 103, 118
Bohr magneton, 4

Class I addition configuration, 127
Class I subtraction configuration, 127

Class II addition configuration, 127–128
Class II subtraction configuration, 128
Closed flux bias, 68
Closed flux pattern, 68
Cobalt dusting, 105–106
CoFe alloys, 106
Coherence length, 43
Coil, 14, 15
Colossal magneto-resistance (CMR) heads, 166–168
Composition, permalloy, 35, 36
Conductor overcoat, 65–66
Core, 14
Critical thicknesses, 103–104
Current
  bias, 51–52
  measuring, 143–144, 152
Current-forcing, voltage-sensing mode, 158

Decay length, 85–87
Demagnetizing fields, 8–11, 12, 29
Density of states, 5
Depth, 142, 151–152
Differential sensing, 126
Digital Compact Cassette (DCC), 59, 62, 73, 124, 134
Digital output pulse, 24–26
Distant sensing, 81
Double-element MRHs
  advantages of, 125, 129–130
  classification of, 126–129
  general comments, 130–131
  output pulse shapes, 126
Double magneto-resistive element bias, 54–55
d shells, 5, 6, 34, 43
Dual-Magneto-Resistor (DMR), 55, 127–128
Dual-Stripe head, 55, 128
Dual stripe synthetic spin valves, 114–125

Dual synthetic spin valves, 114
Dusting of interfaces, 170

EDAC (error detection and correction), 120
Electrical shorting, 121, 122, 125, 129–130
Electromigration phenomenon, 133, 143–144, 155, 160
Electron spin, 3–5
  tunneling junctions, 168–170
Electrostatic discharge (ESD) phenomena, 171–173
Equipartition theorem, 161, 163
Exchange anisotropy, 57
Exchange bias, 56–58
Exchange coupling, 5–6, 101–105
Exchange tabs, 66–67

Faraday's law, 77, 90
Ferrimagnet, synthetic, 113, 155
Ferromagnetism, 5, 6, 155
  free layer and synthetic, 115–116
  pinned layer and synthetic, 111–115
Floating disks, 122, 159
Flux capacity, 115–116
Flux density (B field). *See* Magnetic flux density (B field)
Flux-guide MRHs, 122–123
Fourier transform, 24–25
  of differentiation, 78, 90
Free layers, 94, 96
  synthetic antiferro- and ferrimagnets, 115–116
Free space, 3
Frictional heating effects, 121–122
Fringing field, 12
  recording medium, 21–22

Gap, 14, 15–16
Gap interference, 76

Index

Gap loss, 23
Generalized sinusoid, 74
Giant magneto-resistive (GMR) effect
   discovery of, 39
   granular materials, 47–48
   physics of, 42–46
   superlattices, 39–42
   two-current model, 44–45
Giant magneto-resistive heads (GMRHs)
   attraction of, 94
   spin valve, 94–99
Giant magneto-resistive heads (GMRHs), design of single-element shielded
   future projections, 153–155
   inductive and shielded output pulses, 149–150
   isolated transition peak output voltage, 152–153
   measuring current, 152
   MRE depth, 151–152
   MRE thickness, 150–151
   written magnetization transition, 147–149
Gradiometer heads, 127

Habit, 5
Half-gap, 81
Handshake zap, 171
Hard films, 67
Head field, 14–17
Hewlett-Packard Dual-Stripe head, 55, 128
H field. *See* Magnetic field H
Hilbert transformer, 131
Horizontal bias
   closed flux, 68
   conductor overcoat, 65–66
   exchange tabs, 66–67
   hard films, 67
   purpose of, 64
Hund's rule, 4
Hunt, Robert, 27, 62, 72

unshielded horizontal AMRH, 74–76
unshielded vertical AMRH, 76–78
Hysteresis, 33, 64

IBM, 127, 153
   digital recorders, 73
   disk drives, 91
   tape drives, 60, 91, 124
Induced uniaxial anisotropy, 36
Inductive heads, compared with shielded MRHs
   output pulses, 140–141
   output voltage, 132–136
Inverse Fourier transform, 25
Inverse square law, 1, 8

Johnson/thermal noise, 160, 161–162, 163–164
Joule heating, 143, 144, 155, 171

Karlquist approximation, 90, 99
Keeper effect, 83
Kelvin, Lord, 27
Kodak Dual-Magneto-Resistor (DMR), 55, 127–128

Large-signal voltage comparison, 134–136
Leaks, magnetic flux, 84–87
Lines of force, 12
Lines of sight, 84
Lorentzian pulse shape, 78, 125–126

Magnetic annealing, 102
Magnetic field H, 1–3
Magnetic flux density (B field), 1, 11–12
Magnetic moment μ, 2–3

Magnetic poles, 8–11
Magnetic transmission line, 84–87
Magnetic tunneling junctions (MJTs), 168–170
Magnetization (M field), 1, 7–8
Magnetization fluctuation noise, 161–164
Magnetization-induced process, 70
Magnetization transition, written, 17–20, 138–139, 147–149
Magneto-crystalline anisotropy constant K, 6–7
Magneto-resistive coefficient, 34–35, 133, 159
Magneto-resistive elements (MRE) and sensors, 28–34
   depth, 142, 151–152
   initialization procedures, 68–70
   thickness, 141–142, 150–151
Magneto-resistive heads (MRH)
   See also Double-element MRHs; Single-element shielded vertical MRHs
   disadvantages of, 121
   flux-guide, 122–123
   initialization procedures, 68–70
   yoke-type, 123–124
Magneto-strictive coefficient, permalloy, 36–37
Merged heads, 91
Merging, 91–92
Molecular beam epitaxy (MBE), 39

Noise, 73
   Barkhausen, 34, 64
   double-element MRHs and, 130–131
   Johnson/thermal, 160, 161–162, 163–164
   magnetization fluctuation, 161–164
   signal-to-MRH noise ratios, 159–161

Ohm's law, 29, 75, 161–162
Off-track asymmetry, 79–80, 129
Output pulse
   digital, 24–26
   shapes, 78–79
Output spectrum, 24, 78, 87–91, 99
Output voltage, 23
   comparison of shielded MRHs and inductive heads and, 132–136
   isolated transition peak, 144–146, 152–153

Paramagnet, 5
Peak output voltage, 144–146, 152–153
Penetration length, 43
Permalloy, properties of, 35–38
Permanent magnet bias, 50–51, 65
Permeability μ, 12
Perpendicular bias. *See* Vertical bias
Philips Digital Compact Cassette (DCC), 59, 62, 73, 124, 134
Piggyback heads, 91
Pinned layers, 94, 95
   synthetic antiferro- and ferrimagnets, 111–115
Planck's constant, 4
Pulse amplitude asymmetry, 125, 129
Pulse shapes, single magnetization transition, 78–79
Push-pull operations, 60

Quantum-mechanical exchange phenomenon, 5, 56

Read amplifiers, 157–159
Read-head efficiency, 23
Read-head flux, 22–23
Reading process
   digital output pulse, 24–26
   output spectrum, 24

Index 183

output voltage, 23
read-head flux, 22–23
recording medium fringing fields, 21–22
Reciprocity theorem, 87–89
Ringing down, 70
Ruderman–Kittel–Kasuyu–Yoshida (RKKY) variation, 39, 42, 108, 112, 113, 116

Scaling, magnetic, 155
Self-bias, 55–56, 108–109
Sendust, 38, 91
Servo-bias scheme, 60–62
Shielded anisotropic MRHs. *See* Anisotropic magneto-resistive heads (AMRHs), design of single-element shielded
Shielded giant MRHs. *See* Giant magneto-resistive heads (GMRHs), design of single-element shielded
Shielded magneto-resistive heads (MRHs), compared with inductive heads
  output pulses, 140–141
  output voltage, 132–136
Shielded vertical MRHs. *See* Single-element shielded vertical MRHs
Shields, function of, 81–84
Side-reading asymmetry, 79–80, 125, 129
Signal-to-MRH noise ratios, 159–161
Single-element shielded vertical MRHs
  designs, 91–93
  function of shields, 81–84
  magnetic transmission line, 84–87
  output voltage spectra, 87–91
Small-signal voltage spectral ratio, 132–134
Soft adjacent layer (SAL) bias, 53–54, 86, 92–93, 135, 142
Spacing loss, 23, 75

Spectrum, output, 24, 78, 87–91, 99
Spin, 43
Spin valve giant magneto-resistive heads (GMRHs), 94–99
Spin valves
  biasing effects, 106–109
  cobalt dusting and CoFe alloys, 105–106
  dual stripe synthetic, 114–125
  dual synthetic, 114
  exchange coupling phenomena, 101–105
  resistance change, 98
  top and bottom, 100–101
  unpinning, 102, 109
Split-element scheme, 59–60
Stray field, 12
Streamlines, 12
Subtractive sensing, 126, 127
Superlattices, 39–42

Thermal asperities (TAs), 114, 121, 125, 129
Thermal spikes. *See* Thermal asperities
Thickness, MRE, 141–142, 150–151
Thickness loss, 23, 75
Thomson, William. *See* Kelvin, Lord
Total field, 69
Transition slope parameter, 17, 20, 25
Transverse bias. *See* Vertical bias
Tunneling magneto-resistance ratio (TMR), 169

Unpinning, 102, 109
Unshielded horizontal AMRH, 74–76
Unshielded magneto-resistive (UMR) heads, 78
Unshielded vertical AMRH, 76–78
Upsets, antiferromagnet, 118–120

Vertical bias
 barber-pole scheme, 58–59
 current, 51–52
 double magneto-resistive element, 54–55
 exchange, 56–58
 permanent magnet, 50–51
 self-, 55–56, 108–109
 servo-bias scheme, 60–62
 soft adjacent layer, 53–54, 86, 92–93, 135, 142
 split-element scheme, 59–60
Voltage, output, 23
 comparison of shielded MRHs and inductive heads and, 132–136
 isolated transition peak, 144–146, 152–153
Voltage-forcing, current-sensing mode, 158
Voltage spectral ratio
 large-signal, 134–136
 small-signal, 132–134

Writing process
 coil, 14, 15
 core, 14
 gap, 14, 15–16
 head field, 14–17
 magnetization transition, 17–20, 138–139, 147–149
 yoke, 14

Yoke, 14
Yoke-type MRHs, 123–124, 131

Zero gap length, 78–79